PIC BASIC: Programming and Projects

ME Learn

ampus, Ma

13 8EU

*40   D1141967

PIC BASIC:
Programming and Projects

Dogan Ibrahim

Newnes

AMSTERDAM BOSTON HEIDELBERG LONDON NEW YORK OXFORD
PARIS SAN DIEGO SAN FRANCISCO SINGAPORE SYDNEY TOKYO

Newnes
An imprint of Elsevier
Linacre House, Jordan Hill, Oxford OX2 8DP
30 Corporate Drive, Burlington, MA 01803

First published 2001
Reprinted 2003, 2005

British Library Cataloguing in Publication Data
A catalogue record for this book is available from the British Library

ISBN 0 7506 5229 2

For information on all Newnes publications
visit our web site at www.newnespress.com

Working together to grow
libraries in developing countries

www.elsevier.com | www.bookaid.org | www.sabre.org

ELSEVIER   BOOK AID
            International   Sabre Foundation

Composition by Genesis Typesetting, Laser Quay, Rochester, Kent
Printed and bound in Great Britain by Biddles Ltd,
King's Lynn, Norfolk

Contents

Preface

A microcontroller is a single chip microprocessor system which contains data and program memory, serial and parallel I/O, timers, external and internal interrupts, all integrated into a single chip that can be purchased for as little as $2.00. About 40% of microcontroller applications are in office automation, such as PCs, laser printers, fax machines, intelligent telephones, and so forth. About one-third of micro-controllers are found in consumer electronics goods. Products like CD players, hi-fi equipment, video games, washing machines, cookers and so on fall into this category. The communications market, automotive market, and the military share the rest of the application areas.

Microcontrollers and microprocessors have traditionally been programmed using the assembly language of the target processor. The biggest disadvantage of assembly language is that microcontrollers from different manufacturers have different assembly languages and the user is forced to learn a new language every time a new processor is chosen. Assembly language is also difficult to work with, especially during the development, testing, and maintenance of complex projects. The solution to this problem has been to use a high-level language to program microcontrollers. This approach makes the programs more readable and also portable. The same high-level language can usually be used to program different types of microcontrollers. Testing and the maintenance of microcontroller-based projects are also easier when high-level languages are used.

This book is about programming microcontrollers using a high-level language. The PIC family of microcontrollers is chosen as the target microcontroller. PIC is currently one of the most popular microcontrollers used by many engineers, technicians, students, and hobbyists. PIC microcontrollers are manufactured in different sizes and in varying complexity. These microcontrollers incorporate a reduced instruction set and there is only a small set of instructions that the user has to learn. Also, the power consumption of PIC microcontrollers is very low and this is one of the reasons which made these microcontrollers so popular in portable applications.

BASIC is one of the oldest and the most widely known high-level programming languages. This book is based upon programming the PIC microcontrollers using a BASIC language. The compilers chosen in this book are the PIC BASIC, PIC BASIC PRO and the LET BASIC. The first two compilers are very popular and are developed by MicroEngineering Labs Inc. PIC BASIC is aimed at the lower end of the market, mainly for students and the electronics home market. PIC BASIC PRO is a sophisticated professional compiler with many extra features and this compiler is aimed at engineers and other professional users of PIC microcontrollers. LET BASIC is a cheaper compiler with fewer features and is also aimed at the lower end of the market.

No previous experience with microcontrollers is assumed and the PIC family of microcontrollers is introduced in detail. The book concentrates on the popular 16X84 devices, but examples of other PIC microcontrollers are given where appropriate. PIC BASIC is used in most of the projects but PIC BASIC PRO and LET BASIC are also described where appropriate.

The book is practical and is supplied with many working hardware projects where the reader can experiment easily using a simple breadboard type experiment kit and a few components. The circuit diagram and the code for each project are given where the code is explained in detail.

Chapter 1 provides an introduction to the architecture of the PIC family of microcontrollers. Chapter 2 describes PIC microcontrollers in more detail with special emphasis on the popular 16X84 devices. Chapter 3 describes the features of the PIC BASIC, PIC BASIC PRO and the LET BASIC compilers. PIC development environment and the use of program description language are discussed in Chapter 4. Chapter 5 provides many light projects. The circuit diagrams and the full code listing of each project are given with full comments and explanations. All the projects have been built and tested on a breadboard. Chapter 6 is based on sound projects and there are working projects from simple buzzer circuits to electronic organ projects. Chapter 7 provides several temperature-based projects using digital and analogue temperature sensors. Finally, Chapter 8 describes several RS232-based projects, including the use of serial LCD displays.

Chapter 1

Microcomputer Systems

1.1 Introduction

The term microcomputer is used to describe a system that includes a minimum of a microprocessor, program memory, data memory, and input/output (I/O). Some microcomputer systems include additional components such as timers, counters, analogue-to-digital converters and so on. Thus, a microcomputer system can be anything from a large computer having hard disks, floppy disks and printers, to a single chip computer system.

In this book we are going to consider only the type of microcomputers that consist of a single silicon chip. Such microcomputer systems are also called microcontrollers.

1.2 Microcontroller Systems

Microcontrollers are general purpose microprocessors which have additional parts that allow them to control external devices. Basically, a microcontroller executes a user program which is loaded in its program memory. Under the control of this program data is received from external devices (inputs), manipulated and then sent to external output devices. A microcontroller is a very powerful tool that allows a designer to create sophisticated input/output data manipulation. Microcontrollers are classified by the number of bits in a data word. 8-bit microcontrollers are the most popular ones and are used in many applications. 16- and 32-bit microcontrollers are much more powerful, but usually more expensive and not required in many small to medium general purpose applications where microcontrollers are used.

The simplest microcontroller architecture consists of a microprocessor, memory, and input/output. The microprocessor consists of a central processing unit (CPU) and the control unit (CU).

The CPU is the brain of a microprocessor and is where all of the arithmetic and logical operations are performed. The control unit controls the internal operations of the microprocessor and sends out control signals to other parts of the microprocessor to carry out the required instructions.

Memory is an important part of a microcomputer system. Depending upon the application we can classify memories into two groups: program memory and data

1

memory. Program memory stores all the program code and this memory is usually non-volatile, i.e. data is not lost after the removal of power. Data memory is where the temporary user data is stored during the various arithmetic and logical operations. There are basically five types of memories as summarized below.

1.2.1 RAM

RAM means Random Access Memory. It is a general purpose memory which usually stores user data. RAM is volatile, i.e. data is lost after the removal of power. Most microcontrollers have some amount of internal RAM. 256 bytes is a common amount, although some microcontrollers have more, some less.

1.2.2 ROM

ROM is Read Only Memory. This type of memory usually holds program or fixed user data. ROM memories are programmed at the factory and their contents cannot be changed by the user. ROM memories are only useful if you have developed a program and wish to order several thousand copies of it.

1.2.3 EPROM

EPROM is Erasable Programmable Read Only Memory. This is similar to ROM but the EPROM can be programmed using a suitable programming device. EPROM memories have a small clear window on the chip where the data can be erased under a UV light. Many development versions of microcontrollers are manufactured with EPROM memories where the user program is usually stored. These memories are erased and reprogrammed until the user is satisfied with the program. Some versions of EPROMs, known as OTP (One Time Programmable), can be programmed using a suitable programmer device but these memories cannot be erased. OTP memories cost much less than the EPROMs. OTP is useful after a project has been developed completely and it is required to make hundreds of copies of the program memory.

1.2.4 EEPROM

EEPROM is Electrically Erasable Programmable Read Only Memory. These memories can be erased and also be programmed under program control. EEPROMs are used to save configuration information, maximum and minimum values, identification data etc. Some microcontrollers have built-in EEPROM memories (e.g. PIC16F84 contains a 64-byte EEPROM memory where each byte can be programmed and erased directly by software). EEPROM memories are usually very slow.

1.2.5 Flash EEPROM

This is another version of EEPROM type memory. This type of memory has become popular recently and is used in many microcontrollers (e.g. PIC16F84 contains 1K bytes of flash memory) to store the program data. The data on a flash EEPROM is erased and then reprogrammed using a programming device. The entire contents of the memory should be erased and then reprogrammed. Flash EEPROMs are usually very fast.

One important distinction between a microcontroller and a microprocessor is that a microcontroller has special hardware in the form of input/output ports for dealing with the outside world. Input/output (I/O) ports allow external signals and devices to be connected to the microcontroller. These ports are usually organized into groups of 8 bits and each group is given a name. For example, the PIC16F84 microcontroller contains two I/O ports named port A and port B. It is very common to have at least eight I/O lines. Some microcontrollers have 32 or even 96 I/O lines, where others may have only six. On most microcontrollers the direction of the I/O port lines is programmable so that different bits can be programmed as inputs or outputs. Some microcontrollers provide bi-directional I/O ports where each port line can be used as either input or output. Some microcontrollers have "open-drain" outputs where the output transistors are left floating. External pull-up resistors are normally used with such output port lines.

1.3 Microcontroller Features

Microcontrollers from different manufacturers have different architectures and different capabilities. Some may suit a particular application while others may be totally unsuitable. The hardware features of microcontrollers in general are described in this section.

1.3.1 Supply Voltage

Most microcontrollers operate with the standard +5 V supply. Some micro-controllers can operate at as low as 2.7 V and some will tolerate 6 V without any problems. You should check the manufacturers' data sheets about the allowed limits of the supply voltage.

1.3.2 The Clock

All microcontrollers require an oscillator (known as a clock) to operate. Most microcomputers will operate with a crystal and two capacitors. Some will operate with resonators or with external resistor–capacitor pair. Some microcontrollers have built-in resistor–capacitor type oscillators and they do not require any external timing components (e.g. PIC12C672). If your application is not time sensitive you should use external or internal (if available) resistor–capacitor timing components for simplicity and low cost.

1.3.3 Timers

Timers are an important part of any microcontroller. A timer is basically a counter which is driven from an accurate clock (or a division of this clock). Timers can be

8 bits or 16 bits long. Data can be loaded into the timers and they can be started and stopped under software control. Most timers can be configured to generate an interrupt when they reach a certain count (usually when they overflow). Some microcontrollers offer capture and compare facilities where a timer value can be read when an external event occurs, or the timer value can be compared to a preset value and interrupts can be generated when this value is reached. It is typical to have at least one timer on every microcontroller. Some microcontrollers may have three or more while others may have two timers.

1.3.4 Watchdog

Many microcontrollers have at least one watchdog facility. The watchdog is usually refreshed by the user program and a reset occurs if the program fails to refresh the watchdog. Watchdog facilities are commonly used in real-time systems where it is required to check the proper termination of one or more activities.

1.3.5 Reset Input

This input resets the microcomputer. Most microcontrollers have a resistor connected to the supply voltage and this ensures that the microcontroller starts properly after the application of power. Some microcontrollers have internal reset circuitry which does not require any external components.

1.3.6 Interrupts

Interrupts are a very important concept in microcontrollers. An interrupt causes a microcontroller to respond to external and internal (e.g. timer) events very quickly. When an interrupt occurs the microcontroller leaves its normal flow of execution and jumps directly to the interrupt service routine. Interrupts can in general be nested such that a new interrupt can suspend the execution of another interrupt. Most microcontrollers have at least one, some have several interrupt sources.

1.3.7 Brown-out Detector

Brown-out detectors are also common in many microcontrollers and they reset a microcontroller if the supply voltage falls below a nominal value. Brown-out detectors are usually employed to prevent unpredictable operation at low voltages, especially to protect the contents of EEPROM type memories.

1.3.8 Analogue-to-digital Converter

Some microcontrollers are equipped with analogue-to-digital converter circuits. Usually these converters are 8 bits, but some microcontrollers have 10- or even 12-bit converters. A/D converters usually generate interrupts when a conversion is complete so that the user program can read the converted data very quickly. A/D converters are very useful in control and monitoring applications since most sensors produce analogue output voltages.

1.3.9 Serial Input/Output

Some microcontrollers contain hardware to implement a serial asynchronous communications interface. The baud rate and the data format can usually be selected in software. If serial input/output hardware is not provided, it is easy to develop software to implement serial data transfer using any I/O pin of a microcontroller. Some microcontrollers incorporate SPI (Serial Peripheral Interface) or I²C (Integrated InterConnect) bus interfaces. These enable a microcontroller to interface to other compatible devices easily.

1.3.10 EEPROM Data Memory

EEPROM type memory is also very common in many microcontrollers. The programmer can store non-volatile data in such memory and can also change this data whenever required. Some microcontroller types provide between 64 and 256 bytes of EEPROM data memories, while some others do not have any such memories.

1.3.11 LCD Drivers

LCD drivers enable a microcontroller to be connected to an external LCD display directly. These drivers are not very common since most of the functions provided by them can be implemented by software.

1.3.12 Analogue Comparator

Analogue comparators enable analogue signals to be compared easily. These circuits are not very common and are only implemented in some microcontrollers.

1.3.13 Real-time Clock

The real-time clock is another feature which is implemented in some micro-controllers. These microcontrollers usually keep the date and time of day and they are intended for the consumer market.

1.3.14 Sleep Mode

Some microcontrollers (e.g. PIC) offer sleep modes where executing this instruction puts the microcontroller into a mode where the internal oscillator is stopped and the power consumption is extremely low. The devices usually wake up from the sleep mode by external reset or by a watchdog time-out.

1.3.15 Power-on Reset

Some microcontrollers (e.g. PIC) provide an on-chip power-on reset circuitry which keeps the microcontroller in reset state until all the internal clock and the circuitry are initialized properly.

1.3.16 Low Power Operation

Low power operation is important in portable applications. Some microcontrollers (e.g. PIC) can operate with less than 2 mA with 5 V supply, and around 15 μA at 3 V supply. Some other microcontrollers may consume as much as 80 mA or more at 5 V supply.

1.3.17 Current Sink/Source Capability

This is important if the microcontroller is to be connected to an external device which draws large current for its operation. Some microcontrollers can sink and source only a few mA of current and driver circuits are required if they have to be connected to devices with large current requirements. PIC microcontrollers can sink and source up to 25 mA of current from each I/O pin which is suitable for most small applications, e.g. they can be connected to LEDs without any driver circuits.

1.4 Microcontroller Architectures

Basically, two types of architectures are used in microcontrollers: *Von Neumann* architecture and *Harvard* architecture. Von Neumann architecture is used by a very large percentage of microcontrollers and here all memory space is on the same bus, and instruction and data are treated identically. In the Harvard architecture (used by the PIC microcontrollers), code and data storage are on separate buses and this allows code and data to be fetched simultaneously, resulting in a more efficient implementation.

1.4.1 RISC and CISC

RISC (Reduced Instruction Set Computer) and CISC (Complex Instruction Set Computer) refer to the instruction set of a microcontroller. In a RISC micro-controller, instruction words are more than 8 bits wide (usually 12, 14, or 16 bits) and the instructions occupy one word in the program memory. RISC processors (e.g. PIC) have no more than about 35 instructions, and offer higher speeds. CISC microcontrollers have 8-bit wide instructions and they usually have over 200 instructions. Some instructions (e.g. branch) occupy more than one program memory location.

Chapter 2

The PIC Microcontroller

2.1 The Family

The PIC family of microcontrollers is developed by Microchip Technology Inc. Currently they are one of the most popular microcontrollers, selling over 120 million devices each year. PIC microcontrollers have simple architectures and there are many versions of them, some with only small enhancements and some offering more features. The devices operate on 8-bit wide data and are housed in DIL and SOIC type 8-pin to 64-pin packages.

Basically, all PIC microcontrollers offer the following features:

- RISC instruction set with around 35 instructions

- Digital I/O ports

- On-chip timer with 8-bit prescaler

- Power-on reset

- Watchdog timer

- Power saving SLEEP mode

- Direct, indirect and relative addressing modes

- External clock interface

- RAM data memory

- EPROM (or OTP) program memory

Some devices offer the following additional features:

- Analogue input channels

- Analogue comparators

7

- Additional timer circuits

- EEPROM data memory

- Flash EEPROM program memory

- External and timer interrupts

- Internal oscillator

- USART serial interface

There are basically four families of PIC microcontrollers:

- PIC12CXXX 12/14-bit program word

- PIC16C5X 12-bit program word

- PIC16CXXX 14-bit program word

- PIC17CXXX and PIC18CXXX 16-bit program word

Tables 2.1 to 2.4 give a summary of the features of popular PIC microcontrollers. Some selected devices are described here briefly from each family.

2.1.1 Family PIC12CXXX

PIC12C508: This is a low-cost, 8-pin device with 512×12 EPROM program memory and 25 bytes of RAM data memory. The device can operate at up to 4 MHz clock input and there are only 33 single word instructions. The device features a 6-pin I/O port, 8-bit timer, power-on reset, watchdog timer and internal 4 MHz RC oscillator capability.

The "CE" version of the family (e.g. PIC12CE518) offers an additional 16-byte EEPROM data memory.

Table 2.1 Some PIC12CXXX family members

Microcontroller	Program Memory	Data RAM	Max Speed (MHz)	I/O Ports	A/D Converter
12C508	512×12	25	4	6	–
12C672	2048×14	128	10	6	4
12CE518	512×12	25	4	6	–
12CE673	1024×14	128	10	6	4
12CE674	2048×14	128	10	6	4

Table 2.2   Some PIC16C5X family members

Microcontroller	Program Memory	Data RAM	Max Speed (MHz)	I/O Ports	A/D Converter
16C54	384 × 12	25	20	12	–
16C55	512 × 12	24	20	20	–
16C57	2048 × 12	72	20	20	–
16C58A	2048 × 12	73	20	12	–
16C505	1024 × 12	41	4	12	–

Table 2.3   Some PIC16CXXX family members

Microcontroller	Program Memory	Data RAM	Max Speed (MHz)	I/O Ports	A/D Converter
16C554	512 × 14	80	20	13	–
16C64	2048 × 14	128	20	33	–
16C71	1024 × 14	36	20	13	4
16C77	8192 × 14	368	20	33	8
16F84	1024 × 14	36	10	68	–

Table 2.4   Some PIC17CXXX and PIC18CXXX family members

Microcontroller	Program Memory	Data RAM	Max Speed (MHz)	I/O Ports	A/D Converter
17C43	4096 × 16	454	33	33	–
17C752	8192 × 16	678	33	50	12
18C242	8192 × 16	512	40	23	5
18C252	16384 × 16	1536	40	23	5
18C452	16384 × 16	1536	40	34	8

The high end of this family include devices with 14-bit instruction sets (e.g. PIC12C672) which also have more data RAM and EPROM program memories.

2.1.2 Family PIC16C5X

PIC16C54: This is one of the earliest PIC microcontrollers. The device is 18 pin with a 384 × 12 EPROM program memory, 25 bytes of data RAM, 12 I/O port pins, a timer and a watchdog timer. The device can operate at up to 20 MHz clock input.

Some other members of this family, e.g. PIC16C56, have the same structure but more program memory (1024 × 12). PIC16C58A has more program memory (2048 × 12) and also more data memory (73 bytes of RAM).

2.1.3 Family PIC16CXXX

PIC16C554: This microcontroller has similar architecture to the PIC16C54 but the instructions are 14 bits wide. The program memory is EPROM with 512 × 14 and the data memory is 80 bytes of RAM. There are 13 I/O pins, a timer, and a watchdog timer.

Some other members of this family, e.g. PIC16C71, incorporate four channels of A/D converter, 1024 × 14 EPROM program memory, 36 bytes of data RAM, timer, and watchdog timer. PIC16C77 is a sophisticated microcontroller which offers eight channels of A/D converters, 8192 × 14 program memory, 368 bytes of data memory, 33 I/O pins, USART, I2C bus interface, SPI bus interface, three timers, and a watchdog timer. PIC16F84 is a very popular microcontroller, offering 1024 × 14 flash EEPROM program memory, 36 bytes of data RAM, 64 bytes of EEPROM data memory, 68 I/O port pins, timer, and a watchdog timer.

2.1.4 Family PIC17CXXX and PIC18CXXX

PIC17C42: This microcontroller has a 2048 × 16 program memory. The data memory is 232 bytes. In addition, there are 33 I/O pins, USART, four timers, a watchdog timer, two data capture registers, and pulse width modulator outputs. PIC17C44 is similar but offers more program memory.

PIC18CXXX members of this family include the PIC18C242 type microcontroller with 8192 × 16 program memory, 512 bytes of data memory, 23 I/O pins, five A/D channels (10 bits wide), USART, I^{2}C and SPI bus interfaces, pulse width modulator outputs, four timers, watchdog timer, compare and capture registers, and multiply instructions.

All memory for the PIC microcontroller family is internal and it is usually not very easy to extend the memory externally. No special hardware or software features are provided for extending either the program memory or the data memory. The program memory is usually sufficient for small dedicated projects. But the data memory is generally small and may not be enough for medium to large projects unless a bigger and more expensive member of the family is chosen. For some large projects even this may not be enough and the designer may have to choose a microcontroller from a different manufacturer with a larger data memory, or a microcontroller where the data memory can easily be expanded (e.g. the 8051 series).

2.2 Minimum PIC Configuration

The minimum PIC configuration depends upon the type of microcontroller used, but in general two external parts are needed: reset circuit and oscillator circuit.

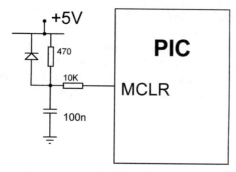

Fig. 2.1 PIC reset circuit

Reset is normally achieved by connecting a 4.7K pull-up resistor from the MCLR input to the supply voltage. Sometimes the voltage rises too slowly and the simple reset function may not work. In this case, the circuit shown in Fig. 2.1 should be used.

PIC microcontrollers have built-in oscillator circuits and this oscillator can be operated in one of five modes:

- LP – Low power crystal

- XT – Crystal/resonator

- HS – High speed crystal/resonator

- RC – Resistor–capacitor

- No external components (only some PICs)

In LP, XT, or HS modes, an external oscillator can be connected to the OSC1 input as shown in Fig. 2.2.

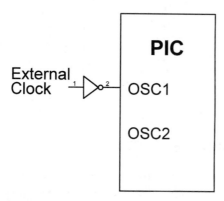

Fig. 2.2 Using an external oscillator

2.2.1 Crystal Operation

As shown in Fig. 2.3, in this mode of operation an external crystal and two capacitors are connected to the OSC1 and OSC2 inputs of the microcontroller. The capacitors should be chosen as in Table 2.5. For example, with a crystal frequency of 4 MHz, two 22 pF capacitors can be used.

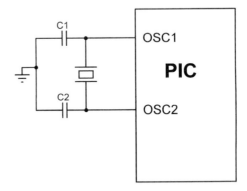

Fig. 2.3 Minimum PIC configuration with a crystal

Table 2.5 Capacitor selection for crystal operation

Mode	Frequency	C1,C2
LP	32 kHz	68–100 pF
LP	200 kHz	15–33 pF
XT	100 kHz	100–150 pF
XT	2 MHz	15–33 pF
XT	4 MHz	15–33 pF
HS	4 MHz	15–33 pF
HS	10 MHz	15–33 pF

2.2.2 Resonator Operation

Resonators are available from 4 MHz to about 8 MHz. They are not as accurate as crystal-based oscillators. Figure 2.4 shows how a resonator can be used with a PIC microcontroller.

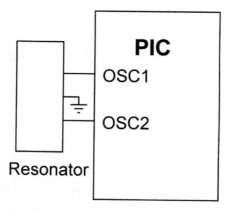

Fig. 2.4 Minimum PIC configuration with a resonator

2.2.3 RC Oscillator

For applications where the timing accuracy is not important we can connect an external resistor and a capacitor to the OSC1 input of the microcontroller as in Fig. 2.5. The oscillator frequency depends upon the values of the resistor and capacitor (see Table 2.6), the supply voltage, and the temperature. For most applications, using a 5K resistor with a 20 pF capacitor gives about 4 MHz and this may be acceptable.

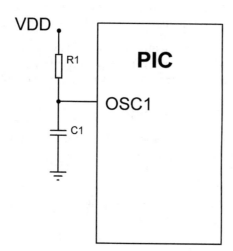

Fig. 2.5 RC oscillator mode

2.2.4 Internal Oscillator

Some PIC microcontrollers (e.g. PIC12C672) have built-in complete oscillator circuits and they do not require any external timing components. The built-in oscillator is usually 4 MHz and is selected during programming of the device.

Table 2.6 RC oscillator component selection

C1	R1	Frequency
20 pF	5 K	4.61 MHz
	10 K	2.66 MHz
	100 K	311 kHz
100 pF	5 K	1.34 MHz
	10 K	756 kHz
	100 K	82.8 kHz
300 pF	5 K	428 kHz
	10 K	243 kHz
	100 K	26.2 kHz

2.3 PIC16F84 Microcontroller

In this section we shall be looking at the architecture of the popular PIC16F84 microcontroller in greater detail. The architectures and instruction sets of other PIC microcontrollers are very similar and with the knowledge gained here we should be able to use any other PIC microcontroller without much difficulty.

Since we shall be programming in BASIC, there is no need to learn the exact details of the architecture or the instruction set. We shall only be looking at the features that a BASIC programmer needs to know while developing software for such devices.

2.3.1 Pin Configuration

Figure 2.6 shows the pin configuration of the PIC16F84. Basically this is an 18-pin device with the following pins:

RB0–RB7   Bi-directional port B pins

RA0–RA4   Bi-directional port A pins

VSS   Ground

VDD   Supply voltage

OSC1   Crystal or resonator (or external clock) input

OSC2   Crystal or resonator input

MCLR   Reset input

INT   This input is shared with RB0 and is also the external interrupt input

TOCK1   This input is shared with RA3 and is also the clock input for the timer

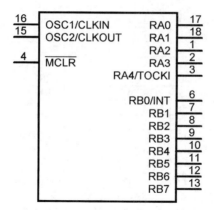

Fig. 2.6 PIC16F84 pin configuration

The microcontroller contains a 1024×14 Flash EEPROM, 68 bytes of data RAM, 64 bytes of EEPROM, 13 I/O pins, timer, and a watchdog timer. Four interrupt sources are available:

● External INT pin interrupt

● Timer overflow interrupt

● Port B (4–7) interrupt on state change

● Data EEPROM write complete

The PIC16F84 normally operates with a 5 V supply and consumes less than 2 mA current at 4 MHz. When operated at 2 V supply it consumes about 15 μA current at 32 kHz. The device is therefore extremely suitable for low-power portable applications.

Register File Map (RFM) is a layout of all the registers available in a microcontroller and this is extremely useful when programming the device, especially when using an assembler language. Figure 2.7 shows the RFM of the PIC16F84 microcontroller. The RFM is divided into two parts: the *Special Function Registers* (SFR), and the *General Purpose Registers* (GPR). On the PIC16F84 device there are 68 GPR registers and these are used to store temporary variables. We shall see later on when programming in BASIC that these registers are used to store the program variables declared by the programmer.

SFR is a collection of registers used by the CPU and peripheral functions to control the internal device operations of the microcontroller. Depending upon the complexity of the devices, some other PIC microcontrollers may have more (or less) SFR registers. It is important that the programmer understands the functions of the SFR registers fully since they are used both in assembly language and in PIC BASIC high-level language.

00	Indirect addr.	Indirect addr.	80
01	TMR0	OPTION	81
02	PCL	PCL	82
03	STATUS	STATUS	83
04	FSR	FSR	84
05	PORTA	TRISA	85
06	PORTB	TRISB	86
07	-	-	87
08	EEDATA	EECON1	88
09	EEADR	EECON2	89
0A	PCLATH	PCLATH	8A
0B	INTCON	INTCON	8B
0C	*68 Bytes General Purpose Registers (0C to 4F)*	*Mapped (accesses in Bank 0)*	8C
...			...
...			...
...			...
4F			CF
50	**50 to 7F Not Implemented**		D0
7F			FF
	BANK 0	**BANK 1**	

Fig. 2.7   Register file map of PIC16F84

The SFR registers used while programming using a high-level language are described in the remainder of this section.

2.3.2 OPTION_REG Register

The OPTION_REG register is a readable and writable register which contains various control bits to configure the on-chip timer and the watchdog timer. This register is at address 81 (hexadecimal) of the microcontroller and its bit definitions are given in Fig. 2.8. For example, to configure the INT pin so that external interrupts are accepted on the rising edge of the INT pin, the following bit pattern should be loaded into the OPTION_REG:

X1XXXXXX

where X is a don't care bit and can be a 0 or a 1.

2.3.3 INTCON Register

This register is readable and writable and contains the various bits for interrupt functions. This register is at address 0B and 8B (hexadecimal) of the microcontroller and the bit definitions are given in Fig. 2.9. For example, to enable interrupts so that

7	6	5	4	3	2	1	0
RBPU	INTEDG	TOCS	TOSE	PSA	PS2	PS1	PS0

Bit 7: PORTB Pull-up Enable
    1 : PORTB pull-ups disabled
    0 : PORTB pull-ups enabled

Bit 6: INT Interrupt Edge Detect
    1 : Interrupt on rising edge of INT input
    0 : Interrupt on falling edge of INT input

Bit 5: TMR0 clock source
    1 : TOCK1 pulse
    0 : Internal instruction cycle

Bit 4: TMR0 Source Edge Select
    1 : Increment on HIGH to LOW of TOCK1
    0 : Increment on LOW to HIGH on TOCK1

Bit 3: Prescaler Assignment
    1 : Prescaler assigned to Watchdog timer
    0 : Prescaler assigned to TMR0

Bit 2-0: Prescaler Rate
    000   1:2
    001   1:4
    010   1:8
    011   1:16
    100   1:32
    101   1:64
    110   1:128
    111   1:256

Fig. 2.8 OPTION_REG bit definitions

external interrupt from pin INT can be accepted the following bit pattern should be loaded into register INTCON:

    1XX1XXXX

2.3.4 TRISA and PORTA Registers

Port A is a 5-bit wide port. Port pins 0 to 3 (i.e. RA0–RA3) have CMOS output drivers. Pin RA4 has an open drain output and should be connected to the supply voltage with a suitable pull-up resistor when used as an output pin. Each port pin has a direction control bit and this bit is stored in register TRISA. Setting a bit in TRISA makes the corresponding PORTA pin an input. Clearing a TRISA bit makes the corresponding PORTA pin an output. For example, to make bits 0 and 1 of port A input and the other bits output, we have to load the TRISA register with:

    00000011

The PORTA register address is 05 and the TRISA address is 85 (hexadecimal).

7	6	5	4	3	2	1	0
GIE	EEIE	TOIE	INTE	RBIE	TOIF	INTF	RBIF

Bit 7: Blobal Interrupt Enable
       1 : Enable all un-masked interrupts
       0 : Disable all interrupts

Bit 6: EE Write Complete Interrupt
       1 : Enable EE write complete interrupt
       0 : Disable EE write complete interrupt

Bit 5: TMR0 Overflow Interrupt
       1 : Enable TMR0 interrupt
       0 : Disable TMR0 interrupt

Bit 4: INT External Interrupt
       1 : Enable INT external interrupt
       0 : Disable INT external interrupt

Bit 3: RB Port Change Interrupt
       1 : Enable RB port change interrupt
       0 : Disable RB port change interrupt

Bit 2: TMR0 Overflow Interrupt Flag
       1 : TMR0 has overflowed (clear in software)
       0 : TMR0 did not overflow

Bit 1: INT Interrupt Flag
       1 : INT interrupt occurred
       0 : INT interrupt did not occur

Bit 0: RB Port Change Interrupt Flag
       1 : One or more of RB4-RB7 pins changed state
       0 : None of RB4-RB7 changed state

Fig. 2.9 INTCON bit definitions

2.3.5 TRISB and PORTB Registers

Port B is an 8-bit wide bi-directional port. The corresponding data direction register is TRISB. A '0' in any TRISB position sets the corresponding port B pins to outputs. A '1' in any TRISB position configures the corresponding port B pins to be inputs. The PORTB register address is at 06 and the TRISB address is 86 (hexadecimal).

Some PIC microcontrollers have more than two ports and these additional ports are named as PORTC, PORTD etc. These ports also have direction registers named as TRISC, TRISD etc.

2.3.6 Timer Module and TMR0 Register

The timer is an 8-bit register (called TMR0) which can be used as a timer or as a counter. When used as a counter, the register increments each time a clock pulse is applied to pin TOCK1 of the microcontroller. When used as a timer, the register increments at a rate determined by the system clock frequency and a prescaler selected by register OPTION_REG. Prescaler rates vary from 1:2 to 1:256. For example, when using a 4 MHz clock, the basic instruction cycle is 1 microsecond (the clock is internally divided by four). If we select a prescaler rate of 1:16, the counter will be incremented at every 16 microseconds.

A timer interrupt is generated when the timer overflows from 255 to 0. This interrupt can be enabled or disabled by bit 5 of the INTCON register. Thus, if we require to generate interrupts at 200 microsecond intervals with a 4 MHz clock, we can select a prescaler value of 1:4 and enable timer interrupts. The timer clock rate is then 4 microseconds. For a time-out of 200 microseconds, we have to send 50 clocks to the timer. Thus, the TMR0 register should be loaded with $256 - 50 = 206$, i.e. a count of 50 before an overflow occurs.

The TMR0 register has the address 01 which can be loaded from either an assembly language or from a high-level language.

The PIC16F84 microcontroller contains 64 bytes of EEPROM data memory. This memory is controlled by registers EEDATA, EEADR and EECON1. There are instructions in the PIC BASIC languages to directly read and write data to this memory and thus these registers will not be discussed here.

The PIC16F84 also contains a *Configuration Register* whose bits can be set or reset during the actual programming of the device. This register includes bits to enable or disable the following features:

- Enable/disable code protection

- Enable/disable power-on timer

- Enable/disable watchdog timer

- Source of the oscillator selection

Some other PIC microcontrollers may have analogue-to-digital converters, pulse width modulator outputs, compare and capture registers and so on. The programming of these features is normally through the SFR register and you should find it easy once you have understood the principles described in this section.

Chapter 3

Using Basic Language to Program PIC Microcontrollers

In this chapter we shall be looking at the principles of programming PIC microcontrollers using a BASIC language. BASIC is one of the oldest and one of the easiest languages to learn. You should be able to learn and program in BASIC in less than an hour.

There are many high-level languages (e.g. BASIC, PASCAL, C etc.) available for the PIC microcontrollers. In this book we shall be looking at how to program our microcontrollers using the PIC BASIC, PIC BASIC PRO and the LET BASIC languages. Our main emphasis will be on the simple and low-cost PIC BASIC language. We shall also have a look at the principles of programming using the other two languages. PIC BASIC is a powerful and yet easy to learn BASIC language. The current version (V1.42) of the PIC BASIC supports most of the 14-bit core PIC microcontrollers, including the PIC16C554, 556, 558, 61, 62, 620, 621, 622, 63, 64, 65, 66, 67, 71, 710, 711, 715, 72, 73, 74, 76, 77, 773, 774, 84, 923, 924, PIC16F83, 84, 873, 874, 876, 877, PIC12C671, 672, and the PIC14C000. PIC BASIC generates code in hex format which can directly be loaded into a microconroller using a suitable programming device. PIC BASIC supports around 40 instructions and a list of these instructions is given in Appendix B.

PIC BASIC PRO is an enhanced version of the PIC BASIC and is aimed mainly at the higher-end professional market. Most of the instructions of the PRO version are similar to the standard version but there are many enhancements in the I/O and in most other areas. The current version (V2.32) of PIC BASIC PRO supports most of the 14-bit and 16-bit core PIC microcontrollers. Over 70 high-level instructions are available and these are listed in Appendix C. PIC BASIC PRO generates code in hex format which can directly be loaded into microcontrollers.

LET BASIC language is aimed at the low end of the market. This language is not as powerful as the PIC BASIC and lacks many important features required in a high-level program. LET BASIC has a limited device support and currently supports only the following PIC microcontrollers: PIC12C508, 509, PIC16C54, 55, 56, 57, 71, 16F83, 16F84, 16F87X. The code generated by LET BASIC is in assembly format of the target microcontroller and this code must be assembled to generate a hex file so that it can be downloaded into the targeted device.

3.1 Variables and the Data RAM

Variables store temporary data in your programs. These variables are stored in the RAM memory of the PIC microcontrollers. Variables are defined and used differently depending upon the type of language used. In this section, we shall be looking at how to store variables using the PIC BASIC, PIC BASIC PRO and the LET BASIC languages.

3.1.1 PIC BASIC Variables

Variables in PIC BASIC can be bytes (8 bits), or words (16 bits). Byte variables are named B0, B1, B2, B3, . . ., and so on. Word variables are named W0, W1, W2, . . . and so forth. Word variables are made up of two bytes. For example, W0 uses the same memory space as bytes B0 and B1. Word variable W1 is made up of bytes B2 and B3, and so on.

Variables are stored in the RAM data memory of PIC microcontrollers. B0 is the first RAM location, B1 is the second RAM location, and so on. The size of the highest available memory location depends upon the microcontroller type. Table 3.1 gives a list of the various PIC microcontrollers and the highest data memory location that should be used.

For example, if we are using a PIC16F84 type microcontroller, we can define 52 variables from B0 to B51, and the highest variable name should not exceed B51. Note that if we try to use higher variables, the compiler does not generate any error messages and the program will not work as expected.

In order to make programs more readable, we can assign meaningful names to variables, instead of using B0, B1 etc. The PIC BASIC statement *symbol* is used for this purpose. For example, we can assign variable name *count* to location B0 using the instruction:

symbol count = B0

Symbols can also be used to assign constants to names. For example, the following statement assigns decimal value 15 to name *maxvalue*. Note that this statement does not occupy any location in the microcontroller data memory. The number is simply represented with a name:

symbol maxvalue = 15

We can access the bit positions of variables B0 and B1 using the predefined name Bit0, Bit1, . . ., Bit15. For example, we can test whether or not bit 1 of variable B0 is set using the statement:

IF Bit1 = 1 THEN . . .

Table 3.1 Highest variable names for various PIC microcontrollers

Microcontroller	Highest byte	Highest word
PIC16C61	B21	W10
PIC16C71	B21	W10
PIC16C710	B21	W10
PIC16F83	B21	W10
PIC16C84	B21	W10
PIC16C711	B51	W25
PIC16F84	B51	W25
PIC16C554	B63	W31
PIC16C556	B63	W31
PIC16C620(A)	B63	W31
PIC16C621(A)	B63	W31
PIC12C67X	B79	W39
PIC14C000	B79	W39
PIC16C558	B79	W39
PIC16C622(A)	B79	W39
PIC16C62(AB)	B79	W39
PIC16C63	B79	W39
PIC16C64(A)	B79	W39
PIC16C65(AB)	B79	W39
PIC16C72(A)	B79	W39
PIC16C73(AB)	B79	W39
PIC16C74(AB)	B79	W39

We can also access the bits of port B using the predefined names Pin0, Pin1, Pin2, . . ., Pin7. Similarly, predefined name *Port* will access all the 8 bits of port B. Table 3.2 gives a list of the PIC BASIC predefined variables.

Numeric constants can be defined in three different ways:

A = 150   assigns decimal value 150 to variable A

A = %00000011   assigns bit pattern "00000011" to variable A

A = $2E   assigns hexadecimal value 2E to variable A

Table 3.2 PIC BASIC predefined variables

Byte	Word	Bit
B0	W0	Bit0 to Bit7
B1		Bit8 to Bit15
B2	W1	
B3		
B4	W2	
B5		
B6	W3	
....	....	
....	....	
B78	W39	
B79		
Pins	Port	Pin0 to Pin7

Character constants are defined by enclosing the character within single quotes. In the following example, ASCII value for decimal 65 is stored in variable M:

M = 'A'

3.1.2 PIC BASIC PRO Variables

Variables in PIC BASIC PRO can be bits (1 bit), bytes (8 bits), or words (16 bits). PIC BASIC PRO, however, is much more sophisticated and allows the programmer to assign names to words, bytes, and bits using the *var* statement and without any predefined memory locations. For example, in the following examples, variables *cnt* and *first* are assigned as bytes, variable *led* is assigned as a single bit:

    cnt   var   byte
    first   var   byte
    led   var   bit

We can also use the var statement to give another name (an alias) for a variable. In the following statement, bit 0 of port B is assigned to variable led:

    led   var   PortB.0

The symbol statement can also be used to give another name for a variable. The statement:

symbol cnt = 10

assigns numeric value 10 to name *cnt*. Note that in this example *cnt* is not a variable and does not occupy a location in memory.

PIC BASIC PRO also supports the assignment of constant values to variables. Constant variables cannot be changed in programs. In the following example, numeric value 100 is assigned to variable *cnt*:

cnt   con   100

PIC BASIC PRO also allows arrays to be created and used in programs. An array is a collection of similar variables. For example, the statement:

first   var   byte[10]

creates 10 bytes and stores them in array *first*. In this example, the elements of the array range from *first[0]* to *first[9]*, yielding 10 elements in total.

Numeric constants can be defined in three different ways:

A = 150   assigns decimal value 150 to variable A

A = %00000011   assigns bit pattern '00000011' to variable A

A = $2E   assigns hexadecimal value 2E to variable A

Character constants are defined by enclosing the character within single quotes. In the following example, ASCII value for decimal 65 is stored in variable M:

M = 'A'

3.1.3 LET BASIC Variables

Variables in LET BASIC are defined using the DIM statement. For example, the following example defines variables *a*, *b*, *first*, and *second*:

DIM a,b,first,second

Predefined name *symbol* is used to assign symbolic names to pins on a port. For example, the statement:

symbol led = B.2

assigns bit 2 of port B to name led.

3.2 PIC BASIC Language Reference

In this section we shall be looking at the PIC BASIC instructions briefly and describe them with simple examples. Detailed information about these instructions can be obtained from the PIC BASIC user guide.

3.2.1 BRANCH

Format: BRANCH offset, (label0, label1, label2, . . .)

Description: This instruction uses byte variable offset to jump to a number of labels. If offset is 0, program jumps to label label0, if offset is 1, program jumps to label1, and so on.

Example: BRANCH B0, (lbl1, lbl2, lbl3)

In this example, if B0 = 0, the program jumps to lbl1

If B0 = 1, the program jumps to lbl2

If B0 = 2, the program jumps to lbl3

3.2.2 BUTTON

Format: BUTTON pin, down, delay, rate, var, action, label

Description: This instruction is used to read the state of a switch connected to a pin (of port B). The parameters are:

pin   pin number (0 to 7).

down   state of the pin when the button is pressed (0 or 1).

delay   delay before auto repeat starts (0 to 255). If delay is 255, switch output is debounced but auto-repeat is not performed.

rate   auto-repeat rate (0 to 255).

var   byte variable used for delay/repeat. This variable should be set to 0 before use.

action   state of button to perform GOTO (0 if switch not pressed, 1 if switch is pressed).

label   Program jumps to this label if action is true.

Example: BUTTON 3, 0, 255, 0, B0, 1, strt

This command checks if the switch connected to pin 3 of port B is pressed. The pin is assumed to be normally pulled high with a resistor. When the switch is pressed the pin goes to 0 (i.e. down = 0). Switch output is debounced. Program jumps to label named strt if the switch is pressed (i.e. action = 1).

3.2.3 CALL

Format: CALL label

Description: Jumps to an assembly language routine named label.

Example: CALL save_data

In the above example, the program jumps to an assembly language routine named *save_data*. The program resumes operation when the return instruction is executed at the end of the called routine.

3.2.4 EEPROM

Format: EEPROM location,constants, . . .

Description: This command stores constants in the on-chip EEPROM data RAM. The command is only available on microcontrollers with EEPROM data RAM (e.g. PIC16F84).

Example: EEPROM 3, (12,10,5)

In this example, bytes 12, 10 and 5 are stored in EEPROM memory locations 3, 4, and 5 respectively.

3.2.5 END

Format: END

Description: This command terminates program memory and enters low-power mode.

Example: END

3.2.6 FOR . . . NEXT

Format: FOR index = start TO end STEP inc

                Statements

      NEXT index

Description: This statement allows repetition in a program. The index can be any variable, byte or word. *Start* defines the first value of index. *End* defines the last value of index. Keyword *STEP* is optional and if present, defines the amount to be added to index at every iteration.

Example: In the example below, the statements within the FOR loop are executed 10 times:

FOR j = 1 TO 10

. . .

. . .

NEXT j

3.2.7 GOSUB

Format: GOSUB label

Description: Program jumps to a subroutine starting at the specified label.

Example: GOSUB led_on

....

led_on: HIGH led   'turn led ON

         PAUSE 1000   'delay 1 second

         LOW led   'turn led OFF

         PAUSE 1000   'delay 1 second

         RETURN   'return from subroutine

In the above example, the program jumps to a subroutine named *led_on* where an led is turned on and off. A RETURN statement should be executed at the end of the subroutine to return to the main program.

3.2.8 GOTO

Format: GOTO label

Description: Program jumps to the statements starting with label.

Example: ....

GOTO loop

....

loop: ....

3.2.9 HIGH

Format: HIGH pin

Description: The specified pin (of port B) is made an output and is set to logic HIGH (+5 V). Pin can take values from 0 to 7.

Example: HIGH 2

In the above example, pin 2 of port B is set to logic HIGH.

3.2.10 I2CIN

Format: I2CIN control, address, var1, var2

Description: This command allows the microcontroller to read data from serial EEPROMs using the I²C bus standard. The high bit of the control indicates whether the address is 8 or 16 bits. If this is 0, then the address is 8 bits. The lower 7 bits of control contain a control code, followed by the chip-select information. The control code for serial EEPROMs is usually '1010'. The I²C data and clock lines are predefined in the PIC BASIC library: pin 0 is the data line and pin 1 is the clock line of port A. A list of some compatible EEPROMs is given below:

Device	Capacity	Control	Address
24LC01B	128 bytes	01010xxx	8 bits
24LC02B	256 bytes	01010xxx	8 bits
24LC32B	4K bytes	11010ddd	16 bits
24LC65	8K bytes	11010ddd	16 bits

where ddd is the device select bits.

Example: symbol control = %01010000

symbol address = B3   'define variable address

address = 5   'set address to 5

I2CIN control, address, B6   'read data from address 5 into B6

3.2.11 I2COUT

Format: I2COUT control, address, value

Description: This command allows one to write data to a serial EEPROM memory using the I²C bus interface. When writing data to a serial EEPROM, there should be about a 10 ms delay for the write command to complete (see I2CIN for the control and address information).

Example: symbol control = %01010000

symbol address = B3

address = 5   'set address to 5

I2COUT control, address, (15)   'write 15 to EEPROM address 5

PAUSE 10   'delay 10 ms

I2COUT control, address, (30)   'write 30 to EEPROM address 5

PAUSE 10   'delay 10 ms

3.2.12 IF . . . THEN

Format: IF condition THEN label

Description: This instruction compares a variable for some condition and if the condition is satisfied then the program jumps to the specified label.

Example: The following code compares the contents of variable x and if x is greater than 10 the program jumps to a label called loop.

IF x > 10 THEN loop

....

....

loop: ....

The following is a list of valid comparisons:

$<$ less than

$\leqq$ less than or equal to

$=$ equal to

$<>$ not equal to

$\geqq$ greater than or equal to

$>$ greater than

In addition to the comparison operators, one can use logical operators AND and OR to combine several conditions.

3.2.13 INPUT

Format: INPUT pin

Description: Makes the specified pin (of port B) an input pin. Pin is a number between 0 and 7.

Example: INPUT 3

The above example makes bit 3 of port B an input pin.

3.2.14 LET

Format: LET var = value operator value

Description: This instruction assigns a value to a variable. This instruction is optional. An operator may be used with another value.

Example: The following code assigns decimal value 27 to variable B1 and also increments B2 by 1:

    LET B1 = 27

    LET B2 = B2 + 1

The following operators are valid:

+	addition
−	subtraction
•	multiplication
**	most significant bit of multiplication
/	division
//	remainder of a division
MIN	minimum
MAX	maximum
&	bitwise AND
|	bitwise OR
^	bitwise Exclusive-OR
&/	bitwise AND NOT
|/	bitwise OR NOT
^/	bitwise XOR NOT

3.2.15 LOOKDOWN

Format: search_value, const_values, return_value

Description: This statement searches a list of constants (*const_values*), comparing each value with a *search_value*. If a match is found, the positional number of the term in the list is stored in variable *return_value*.

Example: In the code:

LOOKDOWN 4, ('12, 45, 56, 4, 5, 2'), B1

The command searches starting from number 12, where the index is 0. A match is found when the index is 3 and thus variable B1 contains the index number 3.

3.2.16 LOOKUP

Format: LOOKUP index, constant_values, value

Description: This statement is used to retrieve data values from a table of constants (constant_values).

Example: In the code below, assume that variable B0 contains number 1:

LOOKUP B0, ("MICRO"), B1

The command retrieves character "I" (index starts from 0) and stores in variable B1.

3.2.17 LOW

Format: LOW pin

Description: Makes the specified output pin (of port B) logic LOW (0 V). Pin must be between 0 and 7.

Example: LOW 2

The above example sets bit 2 of port B to logic LOW.

3.2.18 NAP

Format: NAP period

Description: This command places the PIC microcontroller into low-power mode for the duration of period. Period is given approximately by:

$$Delay = 2^P \times 18\,ms$$

where P is the period. For example, if P is 3, the delay will be 144 ms.

Example: The code:

NAP 1

puts the microcontroller into low-power mode for 36 ms.

3.2.19 OUTPUT

Format: OUTPUT pin

Description: Makes the specified pin (of port B) an output. Pin must be between 0 and 7.

Example: OUTPUT 4

Makes pin 4 of port B an output pin.

3.2.20 PAUSE

Format: PAUSE period

Description: Delays program execution for *period* milliseconds. Period must be between 0 and 65 535.

Example: PAUSE 1000

delays program execution for 1000 ms (1 second).

3.2.21 PEEK

Format: PEEK address, data

Description: This command reads data from a specified register of the micro-controller. Address is the address of the register where data is read.

Example: symbol PORTA = 5   'port A register address

Symbol TRISA = $85   'port address direction register address

POKE TRISA, 255   'make port A pins inputs

PEEK PORTA, B0   'get port A data into B0

The above instruction reads the contents of port A and stores it in variable B0.

3.2.22 POKE

Format: POKE address, value

Description: This instruction writes the *value* to the specified register address.

Example: The instruction POKE $85, 0 writes 0 to register TRISA (address $85) of the microcontroller.

3.2.23 POT

Format: POT pin, scale, variable

Description: This instruction reads a potentiometer (or some other resistive device) on the specified pin. Resistance is measured by timing the discharge of a capacitor through a resistor. Scale must be determined by experiment.

Example: POT 2, 255, B0

The above command reads the potentiometer on pin 2 of port B.

3.2.24 PULSIN

Format: PULSIN pin, state, var

Description: This command measures the pulse width in 10 microsecond increments on the specified pin (of port B). *Var* is a 16-bit number. If *state* is 0, the width of a low pulse is measured. If *state* is 1, the width of the high portion of the pulse is measured.

Example: PULSIN 3, 0, W1

The above command measures the low pulse width on pin 3 of port B, and stores the result in W1.

3.2.25 PULSOUT

Format: PULSOUT pin, period

Description: This instruction generates a pulse on the specified pin (of port B). The period is a 16-bit number that ranges from 0 to 65 535. The pulse width is measured by calculating:

$$\text{Pulse width} = 10\,\mu s \times \text{Period}$$

Example: The command PULSOUT 3, 1000 sends a pulse of 10 000 μs (10 ms) long to pin 3 of port B.

3.2.26 PWM

Format: PWM pin, duty, cycle

Description: This instruction sends a pulse width modulated train of pulses to the specified pin. Each cycle of PWM consists of 256 steps. The duty cycle ranges from 0 to 255.

Example: The command PWM 5, 125, 100 sends a PWM pulse train to pin 5 of port B.

3.2.27 RANDOM

Format: RANDOM var

Description: Generates a random number and stores in *var*. Variable *var* must be 16 bits wide.

Example: RANDOM W1

3.2.28 READ

Format: READ address, var

Description: This instruction reads the EEPROM memory at the specified address and stores the data in *var*. The command is only available for microcontrollers with EEPROM data memory (e.g. PIC16F84).

Example: The instruction READ 3, B0 reads EEPROM location 3 and stores in variable B0.

3.2.29 RETURN

Format: RETURN

Description: Returns from a subroutine.

Example: RETURN

3.2.30 REVERSE

Format: REVERSE pin

Description: The state of the specified pin (of port B) is changed: if the pin is input, it is made an output. If it is output, it is made an input. Pin must be between 0 and 7.

Example: REVERSE 3 changes the state of pin 3 of port B.

3.2.31 SERIN

Format: SERIN pin, mode, qual, item

Description: Allows the microconroller to receive serial RS232 compatible data on the specified pin. Mode sets the baud rate and the data mode (see Chapter 8 for more details).

Example: The command:

SERIN 1, N2400, (B1)

receives serial data with 2400 baud from pin 1 of port B into variable B1.

3.2.32 SEROUT

Format: SEROUT pin, mode, item

Description: Sends serial data out from the specified pin of port B. Mode sets the baud rate and the data mode (see Chapter 8 for more details).

Example: The command:

SEROUT 1, N2400, (B1)

sends serial data in B1, at 2400 baud from pin 1 of port B.

3.2.33 SLEEP

Format: SLEEP period

Description: This command places the microcontroller into low-power mode for a duration of *period* seconds.

Example: The command SLEEP 4 puts the microcontroller into low-power mode for 4 seconds.

3.2.34 SOUND

Format: SOUND pin, note, duration, . . .

Description: This command generates tones and white noise from the specified pin of port B. Note 0 is silence. Notes 1 to 127 are tones, and notes 128 to 255 are white noise (see Chapter 6 for more details). Duration specifies for how long the tune should be played. Each increment of the duration is 12 ms.

Example: The instruction SOUND 1, (100, 10) plays a tune on pin 1 of port B for 120 ms.

3.2.35 TOGGLE

Format: TOGGLE pin

Description: This instruction inverts the state of a pin (of port B). Pin must have a value between 0 and 7.

Example: The instruction TOGGLE 2 changes the state of pin 2 of port B.

3.2.36 WRITE

Format: WRITE address, value

Description: Writes *value* to the specified EEPROM address. This instruction is only valid for microcontrollers which contain on-chip EEPROM data (e.g. PIC16F84).

Example: The instruction:

> WRITE 2, B0

writes the contents of variable B0 to address 2 of the EEPROM data memory.

3.3 PIC BASIC PRO Additional Features

PIC BASIC PRO supports many additional features, some of which are described briefly below. More information can be obtained from the user guide:

● Direct addressing of any port pin, e.g. PORTA.2 addresses bit 2 of port A

● Support for more mathematical functions

● Analogue-to-digital conversion statement (e.g. ADCIN)

● Support of DEBUG statement

● Interrupt support

● Support to produce a frequency output on any pin (e.g. FREQOUT statement)

● Hardware serial I/O support (e.g. HSERIN/HSEROUT statements)

● Complex IF . . . THEN . . . ELSE support

● Standard parallel LCD support (e.g. LCDOUT statement)

- Microsecond delay support (e.g. PAUSEUS statement)

- Serial data I/O support (e.g. SHIFTIN/SHIFTOUT statements)

- WHILE ... WEND support

- XIN-XOUT support

3.4 LET BASIC Additional Features

Some of the additional features of LET BASIC language are described below. More detailed information can be obtained from the user guide:

- Analogue-to-digital conversion statement (ADIN)

- LCD support (e.g. CLS statement)

- BASIC DATA statement support

- Microsecond delay (e.g. DELAYUS)

- IF ... THEN support without changing program flow

- Keypad support

- BASIC INKEY support

- Timer support

Chapter 4

PIC BASIC Project Development

Development of a PIC BASIC project requires several development tools. The following tools are normally required:

- A suitable language compiler. In this book we shall be developing projects using the PIC BASIC, PIC BASIC PRO and the LET BASIC compilers. If you are new to microcontrollers, you may like to choose either the PIC BASIC or the LET BASIC compiler. Otherwise, PIC BASIC PRO is recommended for more professional use.

- A suitable PIC microcontroller programmer device. There are many programmers available on the market for this purpose. The choice here depends upon the amount of money you wish to spend and also the types of microcontrollers you will be programming. For many low end microcontrollers you may like to choose the EPIC programmer, manufactured by the MicroEngineering Labs Inc. This unit is powered from a mains adapter and is connected to a PC with a 25-way parallel port cable. The unit has an on-board 18-pin socket and PIC microcontrollers up to 18 pins can easily be programmed. Adaptors are available to program 28- or 40-pin microcontrollers. Other more expensive programmers may also be chosen which are capable of programming most of the PIC family.

- An EPROM eraser. This is only required if you will be using PIC devices with EPROM windows (e.g. 16C71, 12C672 etc.). For flash microcontrollers (e.g. 16F84) there is no need to purchase an EPROM eraser.

- A PC for program development. This could either be a laptop PC or a standard desktop PC. The CPU speed or the system configuration is not important and most PCs can be used for this purpose.

- A breadboard or some other experimentation kit where you can build and test your projects.

- Depending upon your application, you will also require PIC microcontrollers, crystals, capacitors, resistors, LEDs, LCDs, and other components.

You will normally develop your programs on the PC using a text editor (e.g. Notepad under Windows, or the EDIT command from MSDOS). The program should then be given a meaningful name and saved with an extension '.BAS'. At this stage the program can be compiled using the chosen BASIC compiler. PIC BASIC and PIC BASIC PRO compilers generate '.HEX' files which can directly be loaded into device programmers for programming the microcontroller. LET BASIC generates an assembly file ('.ASM') which should first be assembled using a suitable PIC assembler (e.g. MPASM) to generate the '.HEX' file. This file can then be loaded into a device programmer. If you are using a microcontroller with a flash memory (e.g. 16F84), the program memory on the chip can be erased by the programmer. If, on the other hand, an EPROM type microcontroller is used (e.g. 12C672), an EPROM eraser should be used to erase the program memory before a new program can be loaded onto the chip. Program development is usually an iterative process where you will be erasing and reprogramming the microcontroller many times before you are satisfied with the operation of the program.

In the remainder of this chapter, you will be given programming techniques which should enable you to write structured code which is easier to test and maintain.

4.1 Structure of a Microcontroller-based BASIC Program

The structure of a BASIC program developed for a microcontroller is basically the same as the structure of a standard BASIC program, with a few minor changes. The structure of a typical microcontroller-based BASIC program is shown in Fig. 4.1. It is always advisable to describe the project at the beginning of a program using comment lines. In PIC BASIC and PIC BASIC PRO languages, a single quote is

```
'****************************************************************
'
'          PROJECT:      Give project name
'          FILE:         Give filename
'          DATE:         Date program was developed
'          PROCESSOR:    Give target processor type
'          COMPILER:     Compiler used
'
'
' Describe here what the program does...
'****************************************************************
symbol cnt = ...                     'Enter comments
symbol led = ...                     'Enter comments
symbol sec = ...                     'Enter comments
..........


.........
.........                            'Main body of the program
.........


          END
```

Fig. 4.1   Structure of a microcontroller BASIC program

used to start a comment line. In LET BASIC, the REM statement starts a comment line. The project name, filename, date, and the target processor type should also be included in this part of the program. Any register definition files should then be included for the type of target processor used. This file is supplied as part of the compiler and it includes the definitions for various registers of the microcontroller. The lines of the main program and any subroutines used should also contain comments to clarify the operation of the program.

An example program is shown in Fig. 4.2. This program is developed using the PIC BASIC language. The program turns ON and OFF five times an LED connected to pin 0 of port B of a PIC microcontroller. The program then terminates by performing an endless loop. It is important to realize that there is no returning point in a microcontroller program. Thus, where necessary, an endless loop should be formed at the end to stop the program from going into parts of the program memory which is not programmed.

```
'******************************************************************
'
'           PROJECT:        PROJECT1
'           FILE:           PROJ1.BAS
'           DATE:           August 2000
'           PROCESSOR:      PIC16F84
'           COMPILER:       PIC BASIC
'
'
' This project turns on and off an LED connected to port RB0
' of the microcontroller 5 times.  The flashing rate is 1 second.
'******************************************************************
symbol cnt = B0                     'Assign cnt to location B0
symbol led = 0                      'LED connected to pin 0
symbol sec = 1000                   '1 second delay

        FOR cnt = 1 TO 5            'Do 5 times
            HIGH led                'Turn ON LED connected to RB0
            PAUSE sec               'Delay for 1 second
            LOW led                 'Turn OFF LED connected to RB0
            PAUSE sec               'Delay for 1 second
        NEXT cnt                    'End of loop

loop:   GOTO loop                   'Wait here forever

        END
```

Fig. 4.2 Example microcontroller PIC BASIC program

LET BASIC programs must follow the structure given below (the actual statements used depend upon the program):

DEVICE device

INCLUDE packages

DIM variables

SYMBOL symbols

DEFINE port pins

INIT packages

DATA tables

....

....

....

END

An example LET BASIC program is shown in Fig. 4.3. In this program, an LED connected to pin 3 of port B of the microcontroller is turned ON and OFF 10 times with a delay in between.

```
REM*******************************************************************
REM
REM        PROJECT:        PROJ1
REM        FILE:           PROJ1.BAS
REM        DATE:           August 2000
REM        PROCESSOR:      PIC16F84
REM        COMPILER:       LET BASIC
REM
REM
REM This program turns ON and OFF an LED connected to pin 3
REM of port B 10 times.
REM*******************************************************************

DEVICE 16F84
DIM cnt
SYMBOL led=B.3

FOR cnt = 1 TO 10
        CLEAR led
        DELAYMS(200)
        SET led
        DELAYMS(200)
NEXT cnt

STOP
END
```

Fig. 4.3 Example microcontroller LET BASIC program

4.2 Program Description Language (PDL)

There are many methods that a programmer may choose to describe the algorithm to be implemented by a program. Flowcharts have been used extensively in the past in many computer programming tasks. Although flowcharts are useful, they tend to create unstructured code and also a lot of time is usually wasted drawing them, especially when developing complex programs. In this section we shall be looking at a different way of describing the operation of a program, namely by using a Program Description Language (PDL).

A PDL is an English-like language which can be used to describe the operation of a program. Although there are many variants of PDL, we shall be using simple constructs of PDL in our programming exercises, as described below.

4.2.1 START-END

Every PDL program (or sub-program) should start with a START statement and terminate with an END statement. The keywords in a PDL code should be highlighted in bold to make the code more clear. It is also a good practice to indent program statements between the PDL keywords.

Example:

START

....

....

END

4.2.2 Sequencing

For normal sequencing in a program, write the steps as short English text as if you are describing the program.

Example:

Turn on the valve

Clear the buffer

Turn on the LED

4.2.3 IF-THEN-ELSE-ENDIF

Use IF, THEN, ELSE, and ENDIF statements to describe the conditional flow of control in your programs.

Example:

IF switch = 1 **THEN**

Turn on buzzer

ELSE

Turn off buzzer

Turn off LED

ENDIF

4.2.4 DO-ENDDO

Use DO and ENDDO control statements to show iteration in your PDL code.

Example:

>Turn on LED
>
>**DO** 5 times
>
>>Set clock to 1
>>
>>Set clock to 0
>
>**ENDDO**

Variations of the DO-ENDDO construct are to use other keywords like DO-FOREVER, DO-UNTIL etc. as shown in the following examples.

>Turn off the buzzer
>
>**IF** switch = 1 **THEN**
>
>>**DO UNTIL** Port 1 = 2
>>
>>>Turn on LED
>>>
>>>Read port B
>>
>>**ENDDO**
>
>**ENDIF**

or,

>**DO FOREVER**
>
>>Read data from port B
>>
>>Display data
>>
>>Delay a second
>
>**ENDDO**

4.2.5 REPEAT-UNTIL

This is another useful control construct which can be used in PDL codes. An example is shown below where the program loops until a switch value is equal to 1.

>**REPEAT**
>
>>Turn on buzzer
>>
>>Read switch value
>
>**UNTIL** switch = 1

4.2.6 SELECT

This construct enables us to select an item and do various operations depending upon the value of this item. An example is shown below where if the variable *tmp* is 5, LED is turned ON, if *tmp* is 10, LED is turned OFF, if *tmp* is greater than 10 then a buzzer is turned on:

SELECT tmp

    = 5

        Turn on LED

    = 10

        Turn off LED

    > 10

        Turn on buzzer

END SELECT

PIC BASIC and LET BASIC compilers do not support most of the structured programming constructs and we are forced to use the GOTO statement whenever we wish to change the flow of control in a program. There is no problem using the GOTO statements as long as we use them sparingly and we do not unnecessarily jump back and forward in a program. Too many GOTOs can cause a program to be unreadable and consequently difficult to modify or maintain.

4.3 Internet Websites of PIC Microcontroller Compilers

The amount of microcontroller software available on the Internet is huge and there are many free example programs for a large number of applications, developed by programmers, engineers, hobbyists, students etc. Internet websites of some popular BASIC compilers for the PIC family of microcontrollers, and other useful sites are given below.

4.3.1 BASIC Compilers

PBCSE
http://www.melabs.com

PIC BASIC
http://www.melabs.com
http://www.celestialhorizons.com/products/products.htm
http://www.dontronics.com/piclinks.html

PIC BASIC PRO
http://www.melabs.com

LET BASIC
http://www.crownhill.co.uk

IL-BAS16
http://www.t-online.de/home/il.Stefan.Lehmann/il_hptm2.htm

PIC Basic Pilot
http://www.mal/co.jp/Picbasic/PICbasic.html

PICBAS
http://www.dontronics.com/piclinks.html

FBASIC
http://www.protean-logic.com

FED PIC BASIC
http://www.fored.co.uk/Eindex.htm

4.4 Further Reading

The following books and reference manuals are useful in learning to program the PIC microcontrollers.

Microcontroller Cookbook PIC and 8051, M.R. James, Newnes, ISBN: 0 2405 1448 3

PIC Microcontroller Project Book, J. Iovine, McGraw-Hill, ISBN: 0 07 135479 4

PIC Cookbook, N. Gardner and P. Birnie, Character Press Ltd, ISBN: 1 899013 02 4

50 Things To Do With a PIC, Paul Benfords, Bluebird Technical Press Ltd, ISBN: 1 901631 06 0

Microchip Data On CDROM, Microchip Technology Inc., 2355 W. Chandler Blvd, Chandler, AZ 85224, Website: http://www.microchip.com

PIC BASIC Compiler, MicroEngineering Labs Inc., Box 7532, Colorado Springs CO 80933–7532, Website: http://www.melabs.com

PIC BASIC PRO Compiler, MicroEngineering Labs Inc., Box 7532, Colorado Springs CO 80933–7532, Website: http://www.melabs.com

LET BASIC Compiler, Crownhill Associates Ltd, 32 Broad Street, Ely Cambridgeshire, UK, Website: http://www.crownhill.co.uk

Chapter 5

Light Projects

This chapter describes simple light projects using the basic PIC microcontroller circuits described in early chapters. Most of the projects are based upon the PIC16F84 microcontroller. But the projects and the programs will compile and work on most of the PIC microcontrollers. Over 16 projects are given from very simple LED display projects to complex projects incorporating alphanumeric displays and LCDs. For each project, the following information is given as appropriate:

- *Function*: what the project does, its inputs and outputs.

- *Circuit diagram*: full circuit diagram of the project and explanation of how the circuit works.

- *Program description*: functional description of the software in simple English-like language (PDL).

- *Program listing*: fully tested and working PIC BASIC program listing for each project, including comments. Some projects are based upon the PIC BASIC PRO or the LET BASIC languages.

- *Components required*: listing of components required to build each project.

PROJECT 1 – Flashing LED

Function

This project is very simple and it shows how to turn ON and OFF an LED connected to one of the ports of the microcontroller.

Circuit Diagram

As shown in Fig. 5.1, the circuit is extremely simple, consisting of the basic 16F84 (or any other) PIC microcontroller and an LED connected to bit 0 of port B (i.e. RB0). The manufacturers specify that the maximum sink current of an output pin should not exceed 25 mA. Similarly, the maximum source current of an output pin

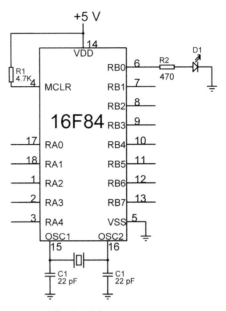

Fig. 5.1   Circuit diagram of Project 1

should be less than or equal to 20 mA. There are many different types of LED lights on the market, emitting red, green, amber, white, or yellow colours. Standard red LEDs require about 5 to 10 mA to emit visible bright light. There are also low-current small LEDs operating from as low as 1 mA. In Fig. 5.1, the microcontroller outputs operate in current sink mode where an LED is turned on if the output port is at logic HIGH level. The required value of the current limiting resistor can be calculated as follows:

$$R = \frac{V_S - V_F}{I_F}$$

where V_S is the supply voltage (+5 V), V_F is the LED forward voltage drop (about 2 V), and I_F is the LED forward current (1 to 30 mA depending on the type of LED used). In this design if we assume an LED current of about 6 mA, the required resistor will be:

$$R = \frac{5 - 2}{6} \cong 470\,\Omega$$

Program Description

The program is required to turn the LED on, then wait for a short while and then turn it off. This process is repeated forever. The following PDL describes the operation of the program:

START

    DO FOREVER

        Turn on the LED

        Wait half a second

        Turn off the LED

        Wait half a second

    ENDDO

END

Program Listing

The full PIC BASIC program listing is shown in Fig. 5.2. Notice that comments are used at the beginning of the program to describe the project title, the filename, the date the project was developed, the target microcontroller chip, and the compiler used. The project, together with its inputs and outputs are then described briefly before the main program.

```
'*************************************************************
'
'          PROJECT:        PROJECT1
'          FILE:           PROJ1.BAS
'          DATE:           August 2000
'          PROCESSOR:      PIC16F84
'          COMPILER:       PIC BASIC
'
'
' This project turns on and off an LED connected to port RB0
' of the microcontroller.  The flashing rate is 1 second.
'*************************************************************
again:  HIGH 0            'Turn on LED connected to RB0
        PAUSE 500         'Delay for .5 seconds

        LOW 0             'Turn off LED connected to RB0
        PAUSE 500         'Delay for .5 seconds

        GOTO again         'Go back and flash the LED forever
        END
```

Fig. 5.2 Program listing of Project 1

HIGH 0 sets the output of RB0 (bit 0 of port RB) to logic level 1 (+5 V), thus turning on the LED. PAUSE 500 delays the program for 0.5 seconds so that we can see the light lit. The LED is then turned off by the command LOW 0. The process is then repeated after a delay of 0.5 seconds by jumping to the beginning of the program using the GOTO command.

The program in Fig. 5.2 can be made more friendly if we assign symbols to the variables used in the program. These are shown in Fig. 5.3 where variable *led* is

```
'*****************************************************************
'
'           PROJECT:          PROJECT1
'           FILE:             PROJ1-1.BAS
'           DATE:             August 2000
'           PROCESSOR:        PIC16F84
'           COMPILER:         PIC BASIC
'
'
' This project turns on and off an LED connected to port RB0
' of the microcontroller.
'*****************************************************************

symbol  led = 0                   'Assign led to 0 (Port RB0)
symbol  second = 1000             'Assign second (1000 ms)

again:  HIGH led                  'Turn LED ON
        PAUSE second              'Delay for 1 second

        LOW led                   'Turn LED OFF
        PAUSE second

        GOTO again                'Go back and flash the LED forever

        END
```

Fig. 5.3 Using symbol assignments in programs

assigned to 0 and variable *second* is assigned to 1000. These variables are then used
in the program, making the program easier to follow.

Figure 5.4 shows how we can use the TOGGLE command to invert the state of an
output pin. This command is only used with port RB and the pins are numbered 0
to 7. In this example, the LED is initially set to off state. The TOGGLE command

```
'*****************************************************************
'
'           PROJECT:          PROJECT1
'           FILE:             PROJ1-2.BAS
'           DATE:             August 2000
'           PROCESSOR:        PIC16F84
'           COMPILER:         PIC BASIC
'
'
' This project turns on and off an LED connected to port RB0
' of the microcontroller 10 times.  The LED is then turned OFF
' permanently.
'*****************************************************************

symbol  led = 0                   'Assign led to 0 (Port RB0)
symbol  second = 1000             'Assign second (1000 ms)
symbol  cnt = B0                  'Assign cnt to B0

        LOW led                   'Turn OFF LED to start with
        FOR cnt = 1 TO 10         'Start of loop
              TOGGLE led          'SET LED ON
              PAUSE second        'Delay for 1 second
              TOGGLE led          'SET LED OFF
              PAUSE second        'Delay for 1 second
        NEXT cnt

        LOW led

        STOP
        END
```

Fig. 5.4 Using the TOGGLE and FOR . . . NEXT commands

changes the state of the LED from off to on and from on to off. The LED is set to flash 10 times. Variable cnt is assigned to byte B0 and a loop is formed using the FOR ... NEXT command. This loop executes 10 times, i.e. the LED flashes 10 times, before it is turned off again.

LET BASIC Code

Figure 5.5 shows how we can program a flashing LED using the LET BASIC language. The device type is defined at the beginning of the program. Port input/output direction is defined using the command:

DEFINE PORTB = 00000000

where all the port B pins are set as outputs. Symbol *LED* is then assigned to pin 0 of port B. The LED is then turned on using the command SET LED and turned off using the command CLEAR LED. Command DELAYMS delays the program so that we can see the LED flashing.

```
rem ************************************************
rem
rem     PROJECT:        PROJECT1
rem     FILE:           PROJ1-3.BAS
rem     DATE:           August 2000
rem     PROCESSOR:      PIC16F84
rem     COMPILER:       LET BASIC
rem
rem
rem This program uses the LET BASIC code
rem The program flashes a LED on and off
rem
rem ************************************************
DEVICE 16F84
DEFINE PORTB=00000000
SYMBOL LED=B.0

LOOP:
        SET LED
        DELAYMS(200)
        CLEAR LED
        DELAYMS(200)
        GOTO LOOP

        END
```

Fig. 5.5 LET BASIC program listing for project 1

Components Required

In addition to the components required by the basic microcontroller circuit, the following components will be required for this project:

R2   470 Ω, 0.125 W resistor

D1   LED

PROJECT 2 – A Lighthouse Flashing LED

Function

This project simulates a marine lighthouse flashing light. An LED is connected to port RB0 of the microcontroller as in Project 1. The program is developed to give three flashes every 10 seconds. The duration of each flash is set to 200 ms. In marine terminology, this lighthouse signal is described as *fl(3) 10*.

Circuit Diagram

The same circuit (Fig. 5.1) as in Project 1 is used. A brighter or a different colour LED can be used if desired.

Program Description

The program is required to turn the LED on and off three times every 10 seconds. A subroutine shall be used to control the LED flashing. The following PDL describes the operation of the program:

Main Program
> **START**
>> **DO FOREVER**
>>> Call subroutine FLASH3 to flash the LED 3 times
>>>
>>> Wait 10 seconds
>> **ENDDO**
> **END**

Subroutine FLASH3
> **START**
>> Turn off LED
>>
>> **DO** 3 times
>>> Turn on LED
>>>
>>> Delay 200 ms
>>>
>>> Turn off LED
>>>
>>> Delay 200 ms
>> **ENDDO**
> **END**

Program Listing

The full PIC BASIC program listing is shown in Fig. 5.6. The main program starts with the label called *again*. Subroutine *FLASH3* is called to flash the LED three times. Then the PAUSE command is used to delay the program for 10 seconds. This process is repeated forever. Inside the subroutine *FLASH3* the LED is turned on and off with a 200 ms duty cycle. A RETURN statement at the end of the subroutine terminates the subroutine and returns control to the main program.

```
'*********************************************************************
'
'       PROJECT:        PROJECT2
'       FILE:           PROJ2.BAS
'       DATE:           August 2000
'       PROCESSOR:      PIC16F84
'       COMPILER:       PIC BASIC
'
'
' This is a simple marine light-house project.  An LED is connected
' to port RB0 of the microcontroller and the LED simulates a
' light-house.  The LED is flashed as follows:
'
'               fl(3)10
'
' i.e. 3 flashes every 10 seconds.
'
'*********************************************************************
symbol  led = 0                 'Assign led to 0 (Port RB0)
symbol  ms200 = 200             'Assign ms200 (200 ms)
symbol  tenseconds = 10000      'Assign 10 seconds (10000ms)
symbol  cnt = B0                'Assign cnt to B0

again:
        GOSUB FLASH3            'Flash 3 times
        PAUSE tenseconds       'Wait for 10 seconds
        GOTO again             'Repeat flashing

'
' This subroutine flashes the LED connected to RB0 3 times
' The interval between the ON and the OFF times is 100ms.
'
FLASH3:
        LOW led                'Turn OFF the LED to start with
        FOR cnt = 1 TO 3       'Start of loop
            HIGH led           'Turn LED ON
            PAUSE ms200        'Delay for 200ms
            LOW led            'Turn LED OFF
            PAUSE ms200        'Delay for 200ms
        NEXT cnt
        RETURN
END
```

Fig. 5.6 Program listing of Project 2

PROJECT 3 – LED Binary Counter

Function

This project counts up in binary and displays the result on eight LEDs connected to port RB of the microcontroller as shown in Fig. 5.7. The project illustrates how multiple outputs can be handled from the BASIC compiler.

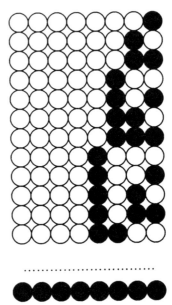

Fig. 5.7 Output pattern of Project 3

Circuit Diagram

As shown in Fig. 5.8 the circuit is very simple, consisting of a 16F84 (or similar microcontroller and eight LEDs connected to port RB of the microcontroller Although eight individual resistors are shown in this circuit, it is possible to replace these resistors with a single DIL (dual-in-line) resistor package.

Program Description

This example shows how multiple outputs can be handled from the BASIC compiler. The LOW and the HIGH commands we have used earlier apply only to a single output of port RB. We can use multiple LOW and HIGH command in order to control more than one output pin. But this approach in general is no very practical and can lead to complex programs which are difficult to understand and maintain.

We should instead use the port registers when it is required to perform multiple output operations in PIC BASIC. The PIC16F84 microcontroller contains two I/C ports, port A and port B. Port A has only five I/O lines. On port B there are eight I/O lines available to the user. Each port has two registers associated with them, the TRIS register and the port data register.

The TRIS register is 8 bits wide and this register controls the direction of data transfer on a particular pin on a port. On power-up and reset, all port A and port B

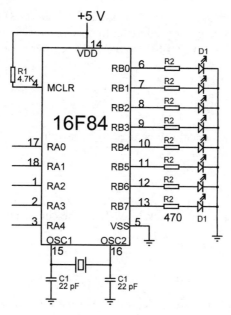

Fig. 5.8 Circuit diagram of Project 3

pins are configured as inputs by default. A bit set to 1 in the TRIS register configures the corresponding port pin to an input. Similarly, a bit cleared in the TRIS register configures the corresponding port pin as an output pin. As an example, if we wish to configure I/O pins 2 and 4 as inputs and the other pins as outputs, the required TRIS bit configuration should be:

Bit positions:	7	6	5	4	3	2	1	0	
Bit weight:	128	64	32	16	8	4	2	1	
Binary value:	0	0	0	1	0	1	0	0	= 20 in decimal

This is equivalent to 14 in hexadecimal, or 20 in decimal and this value should be loaded to the port TRIS register. Similarly, if we wish to configure I/O pins 0, 1, 2 and 3 as outputs and pins 4, 5, 6 and 7 as inputs then the required TRIS configuration is:

Bit positions:	7	6	5	4	3	2	1	0	
Bit weight:	128	64	32	16	8	4	2	1	
Binary value:	1	1	1	1	0	0	0	0	= 240 in decimal

This is equivalent to F0 in hexadecimal, or 240 in decimal. Thus, the port TRIS register should be loaded with this value.

Port A TRIS register has the address 133 decimal (or 85 in hexadecimal). Similarly, the port B TRIS register address is 134 decimal (or 86 in hexadecimal). We can use symbols TRISA and TRISB for the TRIS registers:

    symbol TRISA = 133

    symbol TRISB = 134

The POKE command can be used to write values to the microcontroller registers. The command:

    POKE x,y

loads address x with byte value y. Thus, the command:

    POKE TRISB,10

will load the value 10 to the port B TRIS register.

The actual I/O data is loaded to the port data registers. Port A data register is at address 5, and port B data register is at address 6. We can use the symbol statements to assign values to these registers:

    symbol PORTA = 5

    symbol PORTB = 6

We can then use the POKE command to send data to the ports. As an example, the following command sends data 16 to port B.

    POKE PORTB,16

To summarize, in order to handle multiple outputs on a port, we should first configure the port pins using the TRIS registers and the POKE command. The port data register should then be loaded with the required output data, using again the POKE command.

The following example configures all bits of PORT RB as outputs and then sets bits 0, 1, 2 and 3 (i.e. decimal 15):

    symbol TRISB = 134

    symbol PORTB = 6

    POKE TRISB,0   'configure all bits of port B as outputs

    POKE PORTB,15   'send data 15 (binary 0000 1111) to port B

Instead of specifying the I/O bit configuration as decimal or hexadecimal, it is sometimes easier to use binary notation by making the first character a percent character (i.e. the "%" character). As an example, if we wish to configure port B bit positions 7 and 6 as inputs and the other bit positions as outputs, we could load the TRISB register as follows:

POKE TRISB,%11000000

Table 5.1 gives a summary of the port TRIS and the data register addresses for both port A and port B.

Table 5.1 Port TRIS and data registers

Register	Hexadecimal	Decimal
PORT A	05	5
PORT B	06	6
TRISA	85	133
TRISB	86	134

In this example, we are required to count in binary and display the output pattern on the LEDs as shown in Fig. 5.7. The following PDL describes the operation of the program:

> **START**
>> Define port B TRIS address
>>
>> Define port B data address
>>
>> Configure port B pins as outputs
>>
>> Clear variable LED to 0
>>
>> **DO FOREVER**
>>> Output variable LED to port B
>>>
>>> Delay for 0.5 second
>>>
>>> Increment variable LED by 1
>>
>> **ENDDO**
>
> **END**

Program Listing

The complete program listing is shown in Fig. 5.9. Port B TRIS and the data registers are defined at the beginning of the program. Also, variable *led* is assigned

```
'*****************************************************************
'
'           PROJECT:        PROJECT3
'           FILE:           PROJ3.BAS
'           DATE:           August 2000
'           PROCESSOR:      PIC16F84
'           COMPILER:       PIC BASIC
'
'
' This project counts up in binary and displays the result on
' eight LEDs connected to port RB of the microcontroller.   The
' data is displayed with 0.5 second delay between each output.
'
'*****************************************************************
        symbol TRISB = 134              'Assign TRISB port RB to 134
        symbol PORTB = 6                'Port RB address

        symbol  led = B0                'Assign led to variable B0
        symbol  ms500 = 500             'Assign 0.5 second (500 ms)

        ' Set port RB pins to outputs
                POKE TRISB,0            'Set port RB to output
                led = 0                 'Turn off LED to start with

        again:  POKE PORTB,led          'Output variable led to port RB
                PAUSE ms500             'Wait 0.5 second
                led=led+1               'Increment variable led
                GOTO again              'Go back and repeat

                END
```

Fig. 5.9   Program listing of Project 3

to location B0 and variable *ms500* is defined to have a value of 500. This value will
be used to delay the program by 500 ms. The program then configures all the port
B pins as outputs by using the command:

POKE TRISB,0

Variable *led* is then cleared to zero and the program loop starts. Inside the program
loop, variable *led* is sent to the output port by the command:

POKE PORTB,led

The program is then delayed for 0.5 second so that we can see the LEDs lit, and then
the value of variable *led* is incremented by 1 so that the LEDs display the binary
count. The loop is repeated forever.

Components Required

In addition to the components required by the basic microcontroller circuit, the
following components will be required for this project:

R2   470 Ω, 0.125 W resistor (8 off), or DIL package

D1   LED (8 off)

PROJECT 4 – LED Chasing Project

Function

This project turns on LEDs in sequence connected to port RB of the microcontroller, resulting in a chasing LED effect. The data is displayed with about a 0.5 second delay between each output pattern. Figure 5.10 shows the output pattern displayed by the LEDs.

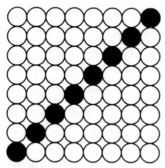

Fig. 5.10 Output pattern of Project 4

Circuit Diagram

The same circuit (Fig. 5.8) as in Project 3 is used. The LEDs can be mounted in a circular or in some other geometric form to enhance the chasing effect. Also, different coloured LEDs can be used to give a colourful output.

Program Description

The program is required to load a 1 into the top (or bottom) bit of a variable and then shift the data right (or left) by one digit and display on the LEDs. A delay will be required between each output. The following PDL describes the function of the program. In this PDL, the bottom bit of a variable is loaded and data is shifted to the left, i.e. multiplied by 2.

> **START**
>
> Define port B TRIS address
>
> Define port B data address
>
> Configure port B pins as outputs
>
> Set count to 1

DO FOREVER

Output count to port RB

Delay 0.5 second

IF count = 128 **THEN**

Set count back to 1

ELSE

Shift count left by 1 digit

ENDIF

ENDDO

END

Program Listing

The full program listing is shown in Fig. 5.11. Variable *led* is used as the count and this variable is initialized to 1 at the beginning of the program. The loop is entered with label *again* and inside this loop the contents of variable *led* are displayed and

```
'****************************************************************
'
'       PROJECT:        PROJECT4
'       FILE:           PROJ4.BAS
'       DATE:           August 2000
'       PROCESSOR:      PIC16F84
'       COMPILER:       PIC BASIC
'
'
' This project turns on the LEDs connected to port RB in sequence.
' Thus, first LED connected to pin RB0 is turned on, then LED on
' RB1, then LED on RB2 and so on until the LED on RB7 is turned on.
' The process is then repeated.  There is a delay of 0.5 seconds
' between each output.
'
'****************************************************************
symbol TRISB = 134              'Assign TRISB port RB to 134
symbol PORTB = 6                'Port RB address

symbol  led = B0                'Assign led to variable B0
symbol  ms500 = 500             'Assign 0.5 second (500 ms)

' Set port RB pins to outputs
        POKE TRISB,0            'Set port RB to output

first:  led = 1                 'Turn on LED 1 to start with
again:  POKE PORTB,led          'Output variable led to port RB
        PAUSE ms500             'Wait 0.5 second
        IF led = 128 THEN first
        led=2*led               'Increment variable led
        GOTO again              'Go back and repeat

        END
```

Fig. 5.11 Program listing of Project 4

the program is delayed for about 0.5 second so that we can see the LEDs lit. The value of variable *led* is then tested and if it is 128 (i.e. the top bit, bit 7, is set) then control is transferred to label *first* where the *led* is reinitialized to 1 for the next round. Otherwise, the value of *led* is shifted left one digit by multiplying by 2. Thus, the values of variable *led* are 1, 2, 4, 8, 16, 32, 64, 128, 1, ... The program loop is repeated forever.

Notice that in this project data is shifted left by multiplying by 2. If we are using the PIC BASIC PRO or the LET BASIC languages then we can shift data left or right using the operators "<<" and ">>" respectively. For example, the code:

B0 = B0 << 4

shifts B0 left four places (same as multiplying by 16). Also, the code:

B0 = B0 >> 2

shifts B0 right by two places (same as multiplying by 4).

PROJECT 5 – Cyclic LED Pattern

Function

This project turns on the LEDs connected to port RB of the microcontroller in a cyclic manner such that first only 1 LED is on, then 2 LEDs are on, then 3, 4, 5, . . ., 8 are on (Fig. 5.12). The process is repeated indefinitely with a 0.5 second delay between each output pattern.

Circuit Diagram

The same circuit (Fig. 5.8) as in Project 3 is used. The LEDs can be mounted in different patterns and in different colours depending upon the application.

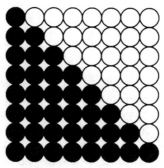

Fig. 5.12 Output pattern of Project 5

Program Description

The program is required to turn the first LED (e.g. corresponding to number 128) and then after a 0.5 second delay, turn on the LED corresponding to numbers 64, 32, 16, and so on until all the LEDs are on (number 255). The process should then be repeated forever as shown in Fig. 5.12 with about a 0.5 second delay between each output pattern. The following PDL describes the functionality of the program:

START

>   Define port B TRIS address
>
>   Define port B data address
>
>   Configure port B pins as outputs
>
>   Set variable count to 128
>
>   Set variable led = count
>
>   **DO FOREVER**
>
>>   Output variable led to port B
>>
>>   Delay 0.5 second
>>
>>   Shift count right by 1 digit
>>
>>   Add count to led
>>
>>   **IF** count = 0 **THEN**
>>
>>>   Set variable count to 128
>>>
>>>   Set variable led = count
>>
>>   **ENDIF**
>
>   **ENDDO**

END

Program Listing

The full program listing is shown in Fig. 5.13. Variable *cnt* is initialized to 128 and variable *led* is set to be equal to *cnt*. The contents of variable *led* are then output to port RB and the program is delayed for 0.5 second so that we can see the LEDs lit. Variable *cnt* is then shifted right by one digit by dividing it by 2. The new value is added to variable *led* so that *led* has the values 128, 128 + 64, 128 + 64 + 32, 128 + 64 + 32 + 16, . . . and so on. When *cnt* reaches 0, it is reinitialized to 128 for the next round.

```
'******************************************************************
'
'           PROJECT:        PROJECT5
'           FILE:           PROJ5.BAS
'           DATE:           August 2000
'           PROCESSOR:      PIC16F84
'           COMPILER:       PIC BASIC
'
'
' This project turns on the LEDs connected to port RB in a cyclic
' manner such that first only 1 LED is on, then 2,3,4,5,...8 are
' all on.  The process is repeated and the data is displayed with
' about 0.5 second delay between each output.
'
'******************************************************************
        symbol TRISB = 134              'Assign TRISB port RB to 134
        symbol PORTB = 6                'Port RB address

        symbol  led = B0                'Assign led to variable B0
        symbol  cnt = B1                'Assign cnt to variable B1
        symbol  ms500 = 500             'Assign 0.5 second (500 ms)

       ' Set port RB pins to outputs
              POKE TRISB,0              'Set port RB to output

       first:  cnt = 128               'Turn on LED 1 to start with
               led = cnt

       again:  POKE PORTB,led           'Output variable led to port RB
               PAUSE ms500              'Wait 0.5 second
               cnt = cnt/2
               led = led + cnt
               IF cnt = 0 THEN first
               GOTO again               'Go back and repeat

               END
```

Fig. 5.13 Program listing of Project 5

Using PIC BASIC PRO and the LET BASIC languages we can shift a data value right using the '>>' shift operator. For example, the command:

cnt = cnt >> 1

produces the same result as the command:

cnt = cnt/2

PROJECT 6 – LED Dice

Function

This project simulates a dice by displaying a random number between 1 and 6, on six LEDs connected to port B of the microcontroller. The LEDs are connected such that they can be turned on and off appropriately to simulate the dots on the faces of a dice. Bit 7 of port B (RB7) is used as the input and a push-button switch is connected to this pin. Every time the switch is pressed, a new number is displayed by the LEDs. The number is displayed for 3 seconds and after this time all the LEDs are turned off, ready for the next switch action.

Circuit Diagram

The circuit diagram of this project is shown in Fig. 5.14. Bit 7 of port B (RB7) is normally held at logic HIGH with a pull-up resistor. When the switch is pressed, RB7 moves to logic LOW and this is detected by the software and a dice number is displayed by the LEDs. As shown in Fig. 5.14, the seven LEDs have been mounted in a pattern to emulate the dots on the faces of a real dice. The pattern displayed for different numbers is shown in Fig. 5.15. As in a real dice, the first row can have up to two LEDs on (corresponding to two dots on a dice), the second row up to three LEDs on, and the third row can have up to two LEDs on.

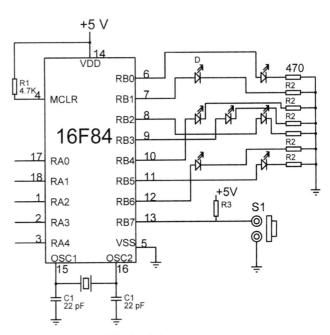

Fig. 5.14 Circuit diagram of Project 6

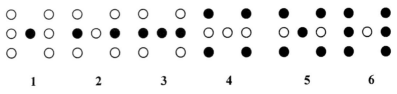

Fig. 5.15 LED pattern displayed for different dice numbers

Program Description

A random number between 1 and 6 shall be obtained during scanning of the push-button switch as follows. The program shall scan the push-button switch continuously. If the switch is not pressed (i.e. at logic HIGH), a number shall be incremented between 1 and 6. Whenever the push-button is pressed, the current value of this number shall be read and this value shall be used as the new dice number. Since the switch is pressed by the user in random, the numbers generated are also random numbers from 1 to 6. The new random number shall be displayed on the seven LEDs appropriately. After about a 3 second delay, all LEDs shall be turned off and the above process shall repeat forever. The following PDL describes the functions of the program:

START

    Define port B TRIS address

    Define port B data address

    Configure RB7 pin as input

    Configure RB0–6 as outputs

    Initialize count to 1

    DO FOREVER

        IF Push-button switch is pressed **THEN**

            Get the new dice number from count

            Turn on the appropriate dice LEDs

            Delay for about 3 seconds

            Turn off all LEDs

        ELSE

            Increment count

            IF count = 7 **THEN**

                Count = 1

            ENDIF

        ENDIF

        ENDDO

    END

Table 5.2 shows the random numbers generated and the corresponding LEDs that will be turned on to give the dice display of Fig. 5.15.

Table 5.2 Dice numbers and corresponding LED patterns

Number	LED ON
1	D4
2	D3,D5
3	D3,D4,D5
4	D1,D2,D6,D7
5	D1,D2,D4,D6,D7
6	D1,D2,D3,D5,D6,D7

```
'*****************************************************************
'
'        PROJECT:          PROJECT6
'        FILE:             PROJ6.BAS
'        DATE:             August 2000
'        PROCESSOR:        PIC16F84
'        COMPILER:         PIC BASIC
'
'
' This project simulates a dice.  Seven LEDs are connected to
' bits 0 to 6 of port RB.  A push-button switch is connected
' to pin 7 of port RB.  The LEDs are configured as a dice face
' as shown below:
'
'        x                 x
'        x        x        x
'        x                 x
'
' When the push-button switch is pressed a dice value is
' displayed by the LEDs.  The dice remains displayed for 3
' seconds and after this time all the LEDs are turned off and
' the process repeats.
'
'*****************************************************************
        symbol TRISB = 134              'Assign TRISB port RB to 134
        symbol PORTB = 6                'Port RB address

        symbol  secs3 = 3000            'Assign 3 seconds (3000 ms)
        symbol  dice  = B0              'Assign dice to location B0

        '
        ' Configure port RB pins
        '
                POKE  TRISB,%10000000   'Set RB0-5 output,RB7 input
                POKE  PORTB,0           'Turn off all LEDs

        start:  dice = 1                'Set dice to 1 to start with
        '
        ' Check switch SW1 status continuously.  If switch is not pressed
        ' then increment the value of dice.  When dice value is 7, it is
        ' reset back to 1.
        ' If the switch is pressed then goto label sw to display
        ' a dice value.
```

Fig. 5.16 Program listing of Project 6

As an example, if the number 3 is to be displayed then only LEDs D3, D4 and D5 will be turned on. Similarly, for number 6, all LEDs except LED 4 will be turned on.

Program Listing

The full program listing is shown in Fig. 5.16. Variable *secs3* is assigned the value 3000 (ms) and this is used to delay the program for 3 seconds. Variable *dice* is assigned to memory location B0. The TRIS register for port B is then configured so

```
loop:     B1=0
          BUTTON 7,0,255,0,B1,1,sw     'Goto label sw if SW1 pressed
          dice=dice+1                  'Increment variable dice
          IF dice = 7 then start       'If 7, reset dice to 1
          goto loop                    'Go and check switch status
      '
      ' The switch is pressed. Now adjust the dice value to be
      ' between 1 and 6 and then branch to a label to display
      ' the value.
      '
      sw:
          dice=dice-1
          BRANCH dice,(n1,n2,n3,n4,n5,n6) 'Branch to display the dice
          goto loop
      '
      ' Below are the 6 labels, n1 to n6 which display a dice value.
      '
      n1:     HIGH 3                   'Display 1
              goto wait

      n2:     HIGH 2                   'Display 2
              HIGH 4
              goto wait

      n3:     HIGH 2                   'Display 3
              HIGH 3
              HIGH 4
              goto wait

      n4:     HIGH 0                   'Display 4
              HIGH 1
              HIGH 5
              HIGH 6
              goto wait

      n5:     HIGH 0                   'Display 5
              HIGH 1
              HIGH 3
              HIGH 5
              HIGH 6
              goto wait

      n6:     HIGH 0                   'Display 6
              HIGH 1
              HIGH 2
              HIGH 4
              HIGH 5
              HIGH 6

      wait:   PAUSE secs3              'Delay 3 seconds
              POKE PORTB,0             'Turn off all LEDs
              GOTO loop                'Go back to repeat

              END
```

Fig. 5.16 (*Continued*)

that RB7 is input and the other port pins are outputs. The value of *dice* is initially set to 1. The status of the push-button switch is then checked using the BUTTON command. The following BUTTON command is used:

BUTTON 7,0,255,0,B0,1,sw

Here, the pin number is specified as 7 (i.e. RB7), the state of this pin when the switch is pressed is 0, the delay is set to 255 so that the command performs a switch debounce action. The auto-repeat rate is set to 0, and variable B0 is used for delay/repeat countdown. The *Action* of the BUTTON command is set to 1 so that a GOTO is performed when the switch is pressed. Finally, the *label* is set to *sw* so that the program execution resumes at label *sw* when the switch is pressed.

If the push-button switch is not pressed, the program continues in sequence and variable *dice* is incremented by 1. When *dice* reaches to 7, it is reset to 1. This process is repeated until the switch is pressed.

```
'****************************************************************
'
'        PROJECT:        PROJECT6
'        FILE:           PROJ6-1.BAS
'        DATE:           August 2000
'        PROCESSOR:      PIC16F84
'        COMPILER:       PIC BASIC
'
'
' This project simulates a dice.  Seven LEDs are connected to
' bits 0 to 6 of port RB.  A push-button switch is connected
' to pin 7 of port RB.  The LEDs are configured as a dice face
' as shown below:
'
'        x                x
'        x        x       x
'        x                x
'
' When the push-button switch is pressed a dice value is
' displayed by the LEDs.  The dice remains displayed for 3
' seconds and after this time all the LEDs are turned off and
' the process repeats.
'
' This code is more efficient than the previous dice code.
'****************************************************************
        symbol TRISB = 134               'Assign TRISB port RB to 134
        symbol PORTB = 6                 'Port RB address

        symbol  secs3 = 3000             'Assign 3 seconds (3000 ms)
        symbol  dice = B0                'Assign dice to location B0
        symbol dice1 = %00001000
        symbol dice2 = %00010100
        symbol dice3 = %00011100
        symbol dice4 = %01100011
        symbol dice5 = %01101011
        symbol dice6 = %01110111


' Configure port RB pins
'
        POKE TRISB,%10000000             'Set RB0-5 output,RB7 input
        POKE PORTB,0                     'Turn off all LEDs
```

Fig. 5.17 More efficient code for Project 6

If the push-button switch is pressed, program execution jumps to label *sw*. Here, the BRANCH command is used to jump to the appropriate routine to turn on the correct LEDs to display the dice number. The BRANCH command is used with the following parameters:

BRANCH dice,(n1,n2,n3,n4,n5,n6)

Here, the *index* is variable *dice*. If *dice* is zero, execution resumes at label *n1*. If *dice* is one, execution resumes at label *n2* and so on. The appropriate LEDs are turned on at labels *n1* to *n6*. For example, at label *n3*, LEDs 2, 3 and 4 are turned on to display the dice number 3. The value of variable *dice* is decremented by one before using the BRANCH command. Thus, the *dice* values are from 0 to 6.

```
start:   dice = 1                              'Set dice to 1 to start with
'
' Check switch SW1 status continuously.  If switch is not pressed
' then increment the value of dice.  When dice value is 7, it is
' reset back to 1.
' If the switch is pressed then goto label sw to display
' a dice value.
'
loop:    B1=0
         BUTTON 7,0,255,0,B1,1,sw              'Goto label sw if SW1 pressed
         dice=dice+1                           'Increment variable dice
         IF dice = 7 then start                'If 7, reset dice to 1
         goto loop                             'Go and check switch status
'
' The switch is pressed. Now adjust the dice value to be
' between 1 and 6 and then branch to a label to display
' the value.
'
sw:
         dice=dice-1
         BRANCH dice, (n1,n2,n3,n4,n5,n6)  'Branch to display the dice
         goto loop
'
' Below are the 6 labels, n1 to n6 which display a dice value.
'
n1:      POKE PORTB,dice1                      'Display 1
         goto wait

n2:      POKE PORTB,dice2                      'Display 2
         goto wait

n3:      POKE PORTB,dice3                      'Display 3
         goto wait

n4:      POKE PORTB,dice4                      'Display 4
         goto wait

n5:      POKE PORTB,dice5                      'Display 5
         goto wait

n6:      POKE PORTB,dice6                      'Display 6

wait:    PAUSE secs3                           'Delay 3 seconds
         POKE PORTB,0                          'Turn off all LEDs
         GOTO loop                             'Go back to repeat

         END
```

Fig. 5.17 *(Continued)*

A More Efficient Program

Notice that in Fig. 5.14, the appropriate LEDs are turned on using the HIGH commands. We can make the program more efficient and easier to understand by using the POKE commands to turn the LEDs on and off. The program listing of a more efficient code is shown in Fig. 5.17. Here, the dice face patterns are assigned to symbols *dice1,dice2, . . . dice6*. The BUTTON command detects when the switch is pressed as before. Also, the BRANCH command sends the program execution to labels *n1* to *n6*, depending upon the dice value. The main difference in this program is the way the dice values are displayed using the POKE commands. The dice face patterns *dice1* to *dice6* are loaded to port B depending upon the value of variable *dice*. The program then waits for 3 seconds and after this time all the LEDs are turned off and control is transferred back to the beginning of the program (i.e. label *loop*).

Programming Using the PIC BASIC PRO Language

This section shows how we can program Project 6 using the PIC BASIC PRO language. The program listing is given in Fig. 5.18. Here, *sec3* is the 3 second delay and it is defined as a constant (*con*). *Dice* is defined as a byte variable. *temp* is a byte variable used by the BUTTON command. The dice faces are stored in a byte array called *dices*. This array has six elements and the first element is *dices[0]*. The binary values of dice faces are stored in six elements of this array. For example, the command:

dices[1] = %00010100

```
'*****************************************************************
'
'        PROJECT:        PROJECT6
'        FILE:           PROJ6-2.BAS
'        DATE:           August 2000
'        PROCESSOR:      PIC16F84
'        COMPILER:       PIC BASIC PRO
'
'
' This project simulates a dice.  Seven LEDs are connected to
' bits 0 to 6 of port RB.  A push-button switch is connected
' to pin 7 of port RB.  The LEDs are configured as a dice face
' as shown below:
'
'
'        x                    x
'        x          x         x
'        x                    x
'
'
' When the push-button switch is pressed a dice value is
' displayed by the LEDs.  The dice remains displayed for 3
' seconds and after this time all the LEDs are turned off and
' the process repeats.
'
' This code is based upon the PIC BASIC PRO compiler and
' demonstrates how arrays can be used in PIC BASIC PRO programs.
' This program will not compile using the standard PIC BASIC
' compiler.
'*****************************************************************
```

Fig. 5.18   PIC BASIC PRO listing for Project 6

stores the binary bit pattern '00010100' in array element *dices[1]*. This element corresponds to dice value 2 and bits 2 and 4 of port B are set to 1.

The program checks the switch status continuously. If the switch is not pressed then the value of *dice* is incremented by 1 and is reset to 1 when it reaches 7. If, on the other hand, the switch is pressed then control is transferred to label *sw*. Here, the value of variable *dice* is decremented so that it is between 0 and 5. This value is then used as an index in array *dices*. The appropriate LEDs are turned on using the command:

PORTB = dices[dice]

The program then waits for 3 seconds and after this time all the LEDs are turned off and control is transferred back to the beginning of the program (i.e. label *loop*).

```
secs3    con      3000              'Constant 3 seconds (3000 ms)
dice     var      byte              'Assign dice to a byte variable
temp     var      byte              'Assign temp to a byte variable

dices    var      byte[6]           'Assign dices to 6 array bytes

dices[0]=%00001000                  'Dice 1
dices[1]=%00010100                  'Dice 2
dices[2]=%00011100                  'Dice 3
dices[3]=%01100011                  'Dice 4
dices[4]=%01101011                  'Dice 5
dices[5]=%01110111                  'Dice 6
'
' Configure port RB pins
'
         TRISB=%10000000            'Set RB0-5 output,RB7 input
         PORTB=0                    'Turn off all LEDs

         dice = 1                   'Set dice to 1 to start with
'
' Check switch status continuously.  If switch is not pressed
' then increment the value of dice.  When dice value is 7, it
' is reset back to 1.
' If the switch is pressed then goto label sw to display
' a dice value.
'
loop:    temp=0
         BUTTON PORTB.7,0,255,0,temp,1,sw  'Goto label sw if SW1 pressed
         dice=dice+1                       'Increment variable dice
         IF dice = 7 then                  'If 7, reset dice to 1
            dice=1
         ENDIF
         goto loop                         'Go and check switch status
'
' The switch is pressed. Now adjust the dice value to be
' between 1 and 6 and then display the value.
'
sw:
         dice=dice-1               'dice between 1 and 6
         PORTB=dices[dice]         'Display dice value
         PAUSE secs3               'Delay 3 seconds
         PORTB=0                   'Turn off all LEDs
         goto loop                 'Go back and repeat

         END
```

Fig. 5.18 (*Continued*)

It is clear from the program listing that the PIC BASIC PRO language is a more compact and more powerful language, supporting most features of the BASIC language found on PCs and big computer systems.

Components Required

In addition to the components required by the basic microcontroller circuit, the following components will be required for this project:

R2　470 Ω, 0.125 W resistor (7 off)

R3　100 K, 0.125 W resistor

D1　LEDs (7 off)

S1　Push-button switch

PROJECT 7 – Hexadecimal Display

Function

This project shows how a PIC microcontroller can be interfaced to a TIL311 type hexadecimal display. The program counts up from 0 to 9 and then in hexadecimal format from A to F and then back to 0. This process is repeated forever with a 0.5 second delay inserted between each count.

Circuit Diagram

The circuit diagram of this project is shown in Fig. 5.19. TIL311 is a popular 14-pin DIL small hexadecimal display, powered from a +5 V supply. Inputs A, B, C, D of the display are the data inputs and these are connected to the lower part of port B (RB0, RB1, RB2 and RB3). LATCH Input (pin 5) controls the display. When LATCH is LOW, new data is written to the display. When LATCH is HIGH, the display data is frozen. A new data is displayed by sending the data to the A, B, C, D inputs and then the LATCH input is set to logic LOW and then back to logic HIGH (i.e. pulsed). The data write sequence is thus as follows:

● Make sure the LATCH is at logic HIGH

● Send 4-bit data to the display

● Set LATCH to logic LOW

● Set LATCH to logic HIGH

The LATCH input of the display is connected to bit 7 of port B (RB7). Pins 1 and 14 of the display are connected to +5 V and pins 7 and 8 are connected to the ground.

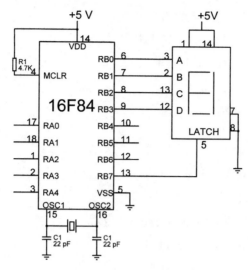

Fig. 5.19 Circuit diagram of Project 7

Program Description

The program is very simple. The count is initially set to 0 and the display latch is set to 1 to avoid any unintentional write to the display. The count is then sent to the display and the display latch is clocked. The next data value is obtained by incrementing the count. When the count reachs 16, it is reset to 0. The following PDL describes the functions of the program. Subroutine TIL311 displays the contents of variable *cnt* on the TIL311:

Main program

    START

        Define port B TRIS address

        Define port B data address

        Define latch as bit 7

        Configure port B pins as outputs

        Set display latch to 1

        Set cnt to 0

    DO FOREVER

        Call subroutine TIL311 to display cnt

        Delay 0.5 second

>>   Increment cnt
>>   **IF** cnt = 16 **THEN**
>>   cnt = 0
>>   **ENDIF**
>>   **ENDDO**
> **END**

Subroutine TIL311

> **START**
>>   Set display latch and output cnt
>>   Set LATCH to LOW
>>   Set LATCH to HIGH
>>   Return
> **END**

Program Listing

The full program listing is shown in Fig. 5.20. Variable *cnt* is assigned to memory location B0, *disp* is assigned to memory location B1 and *latch* is assigned to number 7 which will be used to manipulate bit 7 of port B. Port B pins are then configured as outputs and the display latch is set to HIGH to prevent any unintentional write to the display. Variable *cnt* is then set to 0 and subroutine TIL311 is called to display the value of *cnt*. After a 0.5 second delay the value of *cnt* is incremented and the process is repeated. When *cnt* reaches 16, it is reset to 0 by jumping to program label *strt*. The displayed data is:

    0 1 2 3 4 5 6 7 8 9 A B C D E F 0 1 . . .

Subroutine TIL311 displays the contents of variable *cnt*. The latch is initially set to 1 by adding 128 (i.e. setting the top bit) to *cnt*. The POKE command sends the value of *cnt* to port B of the microcontroller. The display latch is then cleared to 0 to enable the data to be written to the display and then back to 1 to freeze the display.

Components Required

In addition to the components required by the basic microcontroller circuit, a TIL311 type alphanumeric display will be required.

```
'*********************************************************************
'
'           PROJECT:        PROJECT7
'           FILE:           PROJ7.BAS
'           DATE:           August 2000
'           PROCESSOR:      PIC16F84
'           COMPILER:       PIC BASIC
'
'
' This is a simple counter project.  A TIL311 type hexadecimal
' alphanumeric display is connected to port RB of the microcontroller.
' The program counts from 0 to 9 and then from A to F in hexadecimal.
' The data is displayed with about 0.5 second delay between each
' output value.
'
'*********************************************************************
       symbol TRISB = 134               'Assign TRISB port RB
       symbol PORTB = 6                 'Assign RB address
       symbol  ms500 = 500              'Assign 500 ms
       symbol  cnt = B0                 'Assign cnt to B0
       symbol  disp = B1                'Assign disp to B1
       symbol  latch = 7                'Latch is on bit 7 of port B

'
' Configure port RB pins
'
             POKE TRISB,0               'Set Port RB pins to outputs
             HIGH latch                 'SET latch HIGH to start with

strt:   cnt = 0                         'Clear cnt to start with
loop:   GOSUB TIL311                    'Display the value of cnt
             PAUSE ms500                'Delay 0.5 second
             cnt=cnt+1                  'Increment cnt
             IF cnt = 16 then strt      'If 16, reset to 0
             GOTO loop                  'Repeat the counting


'
' This subroutine displays the contents of variable cnt on
' the TIL311 hexadecimal alphanumeric display.
'
TIL311:
             disp=128+cnt               'Set top bit of data and load into disp
             POKE PORTB,disp            'Send data to port RB
             LOW latch                  'Clear latch
             HIGH latch                 'Set latch
             RETURN                     'Return to main program
END
```

Fig. 5.20   Program listing of Project 7

PROJECT 8 – Two-digit Decimal Count

Function

This project shows how a PIC microcontroller can be interfaced to two TIL311 type hexadecimal displays. The program counts up continuously from 0 to 99 in decimal with about a 0.5 second delay between each count.

Circuit Diagram

The circuit diagram of this project is shown in Fig. 5.21. Two TIL311 type hexadecimal displays are used. Display MSD (Most Significant Digit) will be

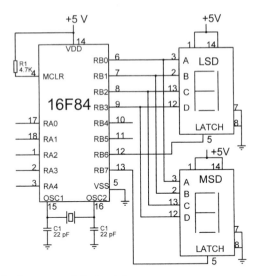

Fig. 5.21   Circuit diagram of Project 8

programmed to show the tens and the LSD (Least Significant Digit) will show the units. Data inputs (A, B, C, D) of both displays are connected in parallel to the lower part of port B (RB0 to RB3). Latch inputs (pin 5) of the displays are controlled separately. Latch input of display MSD is connected to port pin RB7 and the same input of display LSD is connected to port pin RB6 of the microcontroller.

MSD data is displayed by sending the data to port B and then clocking pin RB7. Similarly, LSD data is displayed by sending data to port B, but this time pin RB6 is clocked.

Program Description

Port B pins are first set as outputs and count is set to 0. Subroutine TIL311_2 is then called to display the value of count as two decimal digits. This subroutine separates the count into two decimal parts, MSD and LSD, and sends each part to the appropriate display. The next display value is obtained by incrementing the count. When count reaches 100, it is reset to 0 and the process is repeated forever. The following PDL describes the functions of the program.

Main Program

    START

            Define port B TRIS address

            Define port B data address

            Define latch_msd as bit 7

Define latch_lsd as bit 6

Configure port B pins as outputs

Set both display latches to 1

Set cnt to 0

DO FOREVER

> Call subroutine TIL311_2 to display cnt
>
> Delay 0.5 second
>
> Increment cnt
>
> **IF** cnt = 100 **THEN**
>
> > cnt = 0
>
> **ENDIF**
>
> **ENDDO**

END

Subroutine TIL311_2

> **START**
>
> > Extract the MSD digit of cnt
> >
> > Extract the LSD digit of cnt
> >
> > Set both latches HIGH with MSD data
> >
> > Clear the MSD latch to 0
> >
> > Set the MSD latch to 1
> >
> > Set both latches HIGH with LSD data
> >
> > Clear the LSD latch to 0
> >
> > Set the LSD latch to 1
> >
> > Return
>
> **END**

Program Listing

The full program listing is shown in Fig. 5.22. Latch input of display MSD is named *latch_msd* and is assigned to 7. Similarly, latch input of display LSD is named *latch_lsd* and is assigned to 6. The count (variable *cnt*) is initially set to

```
'*********************************************************************
'
'           PROJECT:        PROJECT8
'           FILE:           PROJ8.BAS
'           DATE:           August 2000
'           PROCESSOR:      PIC16F84
'           COMPILER:       PIC BASIC
'
'
' This is a simple 2 digit counter project.  Two TIL311 type hexadecimal
' alphanumeric displays are connected to port RB of the microcontroller.
' The program counts continuously from 0 to 99 in decimal.
' The data is displayed with about 0.5 second delay between each
' output value.
'
'*********************************************************************
        symbol TRISB = 134              'Assign TRISB port RB
        symbol PORTB = 6                'Assign RB address
        symbol  ms500 = 500             'Assign 500 ms
        symbol  cnt = B0                'Assign cnt to B0
        symbol  disp = B1               'Assign disp to B1
        symbol  msd = B2                'Assign MSD memory location
        symbol  lsd = B3                'Assign LSD memory location
        symbol  latch_msd = 7           'Latch msd is on bit 7 of port B
        symbol  latch_lsd = 6           'Latch lsd is on bit 6 of port B


    '
    ' Configure port RB pins
    '
            POKE TRISB,0                'Set Port RB pins to outputs
            HIGH latch_msd              'Set latch msd HIGH to start with
            HIGH latch_lsd              'Set latch lsd HIGH to start with

strt:       cnt = 0                     'Clear cnt to start with
loop:       GOSUB TIL311_2              'Display the value of cnt
            PAUSE ms500                 'Delay 0.5 second
            cnt=cnt+1                   'Increment cnt
            IF cnt = 100 then strt      'If 100, reset to 0
            GOTO loop                   'Repeat the counting


    '
    ' This subroutine displays the contents of variable cnt on
    ' two TIL311 type hexadecimal alphanumeric display.  The display
    ' on the left (tens) is named MSD and the display on the right
    ' (units) is named LSD.
    '
TIL311_2:
            msd=cnt/10                  'Extract the MSD digit
            lsd=cnt//10                 'Extract the LSD digit (remainder)
            disp=128+64+msd             'Set both latches HIGH with msd data
            POKE PORTB,disp             'Send msd data to the displays
            LOW latch_msd               'Clear msd latch
            HIGH latch_msd              'Set msd latch
            disp=128+64+lsd             'Set both latches HIGH with lsd data
            POKE PORTB,disp             'Send lsd data to the displays
            LOW latch_lsd               'Clear lsd latch
            HIGH latch_lsd              'Set lsd latch
            RETURN                      'Return to main program
END
```

Fig. 5.22 Program listing of Project 8

0 and both latches are set to 1 to avoid any accidental write to the displays. An endless loop is then formed and subroutine TIL311_2 is called to display the value of variable *cnt*. The count is then incremented by 1 and when it reaches 100, it is reset to 0. The loop is repeated with a 0.5 second delay between each output value. The displayed data is:

0 1 2 3 4 5 6 7 8 9 10 11 ... 98 99 0 1 ...

Subroutine TIL311_2 displays the contents of variable *cnt* as two decimal digits. The MSD part of *cnt* is obtained by dividing *cnt* by 10. The LSD part of *cnt* is obtained by taking the remainder after a division by 10. The following command stores the remainder of *cnt* in variable *lsd* after dividing it by 10:

lsd = cnt // 10

The MSD digit is displayed by sending the MSD data to the display and clocking the *latch_msd*. Similarly, the LSD digit is displayed by sending the LSD data to the display and then clocking the *latch_lsd*.

Programming Using the PIC BASIC PRO Language

This section shows how we can program Project 8 using the PIC BASIC PRO language. The full program listing is given in Fig. 5.23. Here, variables *cnt, msd* and *lsd* are assigned to memory locations using the *var* statements. *cnt* is the actual count displayed, *msd* stores the MSD digit of the count and *lsd* stores the LSD digit of the count. Port B pins are set as outputs using the command TRISB = 0. The two display latches are then set to logic HIGH using the commands:

latch_msd = 1

latch_lsd = 1

The count is initialized to 0 (*cnt = 0*) and subroutine TIL311_2 is used to display the value of *cnt*. After a 0.5 second display the value *cnt* is incremented by 1 and the whole process is repeated. When *cnt* reaches 100, it is reset to 0.

Inside the subroutine TIL311_2, the MSD and the LSD digits are extracted using the following commands:

msd = cnt / 10

lsd = cnt // 10

After the display data is sent out to port B, the PULSOUT command is used to clock the display latches. Initially a display latch is at logic HIGH. The PULSOUT command will clear the latch to LOW and then back to HIGH, i.e. a clock pulse will be generated on the specified output pin. The duration of this pulse is not very

```
'****************************************************************
'
'        PROJECT:        PROJECT8
'        FILE:           PROJ8-1.BAS
'        DATE:           August 2000
'        PROCESSOR:      PIC16F84
'        COMPILER:       PIC BASIC PRO
'
'
' This is a simple 2 digit counter project.  Two TIL311 type hexadecimal
' alphanumeric displays are connected to port RB of the microcontroller.
' The program counts continuously from 0 to 99 in decimal.
' The data is displayed with about 0.5 second delay between each
' output value.
'
' This program is only for the PIC BASIC PRO version and it will
' not compile using the standard PIC BASIC compiler.
'
'****************************************************************
cnt       var      byte             'Assign cnt
msd       var      byte             'Assign MSD
lsd       var      byte             'Assign LSD
symbol  ms500 = 500                 'Assign ms500 to 500 ms delay
latch_msd var     PORTB.7           'Rename PORTB.7 as latch_msd
latch_lsd var     PORTB.6           'Rename PORTB.6 as latch_lsd


'
' Configure port RB pins
'
          TRISB = 0                 'Set Port RB pins to outputs
          latch_msd = 1             'Set latch_msd
          latch_lsd = 1             'Set latch_lsd

strt:     cnt = 0                   'Clear cnt to start with
loop:     GOSUB TIL311_2            'Display the value of cnt
          PAUSE ms500               'Delay 0.5 second
          cnt = cnt+1               'Increment cnt
          IF cnt = 100 then         'If 100, reset to 0
             cnt = 0
          ENDIF
          GOTO loop                 'Repeat the counting


'
' This subroutine displays the contents of variable cnt on
' two TIL311 type hexadecimal alphanumeric display.  The display
' on the left (tens) is named MSD and the display on the right
' (units) is named LSD.
'
TIL311_2:
          msd = cnt/10              'Extract the MSD digit
          lsd = cnt//10             'Extract the LSD digit (remainder)
          PORTB = 128+64+msd        'Set both latches HIGH with msd data
          PULSOUT latch_msd,1       'Send a pulse to msd latch
          PORTB = 128+64+lsd        'Send lsd data to the displays
          PULSOUT latch_lsd,1       'Send a pulse to lsd latch
          RETURN                    'Return to main program
END
```

Fig. 5.23 PIC BASIC PRO listing for Project 8

critical in this project and is set to 1 period (1 period is equivalent to 10 μs with a 4 MHz clock frequency) using the command:

PULSOUT latch_msd,1

and

PULSOUT latch_lsd,1

Components Required

In addition to the components required by the basic microcontroller circuit, two TIL311 type alphanumeric displays will be required.

PROJECT 9 – TIL311 Dice

Function

This project is a dice, constructed from a TIL311 type hexadecimal display. When a push-button switch, connected to bit 7 of port B, is depressed, a random number between 1 and 6 is displayed on the display. After about 2 seconds, the display is cleared and the user can throw a dice again. The program runs in an endless loop.

This project demonstrates how we can use the PEEK command to read data from an I/O port.

Circuit Diagram

The circuit diagram of this project is shown in Fig. 5.24. A TIL311 type hexadecimal display is connected as in Project 7. Additionally, a push-button switch is connected to bit 7 of port B (RB7). This pin is normally held at logic 1 with a pull-up resistor, and goes to logic 0 when the switch is pressed.

Program Description

The display latch is initially set to logic 1 to avoid any accidental data display. The state of the push-button switch is then checked continuously and when the button is not pressed, a count is incremented between 1 and 6. When the push-button switch is pressed, the current value of the count is sent to the display by calling subroutine TIL311. The above process continues after about 3 seconds of delay between each dice display.

The following PDL describes the functions of the program:

Main Program

    START

        Define port B TRIS address

        Define port B data address

        Configure RB0–6 as outputs, RB7 as input

        Clear display

        Set dice value to 1

        DO FOREVER

            IF switch pressed **THEN**

                Call subroutine TIL311 to display new dice value

                Delay 3 seconds

            ELSE

                Increment dice value

                IF dice value = 7 **THEN**

                    Dice value = 1

                ENDIF

            ENDIF

        ENDDO

    END

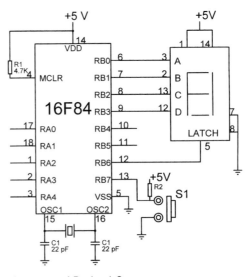

Fig. 5.24　Circuit diagram of Project 9

Subroutine TIL311

    START

        Set display latch

        Send data to the display

        Clear display latch to 0

        Set display latch to 1

        Return

    END

Program Listing

The full program listing is shown in Fig. 5.25. The display latch is assigned to number 6, *secs3* is assigned to number 3000 which is used to delay the program by 3 seconds, variable *dice* is allocated the memory location B1 and variable *disp* is allocated the memory location B1. The POKE command:

    POKE TRISB,%10000000

is used to set bits 0–6 of port B as outputs and bit 7 as an input pin. The display latch is set to 1 and then the display is cleared at the beginning of the program by using the command:

    POKE PORTB,0

The switch status is then checked continuously. Pin RB7 is normally held at logic HIGH by using a pull-up resistor. When the switch is pressed, pin RB7 is lowered to logic LOW and this is detected by the program using the commands:

    PEEK PORTB,B0

    IF BIT7 = 0 THEN sw

The PEEK command reads the status of port B pins into variable B0. Remember that the bits of B0 can be accessed as BIT0, BIT1, . . ., BIT7. The program checks the status of bit 7 of port B and then jumps to label *sw* if this bit is 0 (i.e. at logic LOW). Otherwise, the value of *dice* is incremented by 1 and is reset to 1 when it reaches 7. At label *sw*, the program calls to subroutine TIL311 to display the dice value. The program then delays for 3 seconds and jumps back to the beginning of the *loop*, ready for a new dice.

Components Required

In addition to the standard components, the following components will be required for this project:

    TIL311 display

    Push-button switch

```
'*****************************************************************
'
'          PROJECT:        PROJECT9
'          FILE:           PROJ9.BAS
'          DATE:           August 2000
'          PROCESSOR:      PIC16F84
'          COMPILER:       PIC BASIC
'
'
' This is a simple dice simulator project.
'
' A TIL311 type hexadecimal alphanumeric display is connected
' to port B of the PIC microcontroller.  When a push-button
' switch, connected to bit 7 of port B is pressed, a random
' number is displayed between 1 and 6.  After about 3 seconds
' delay the display is cleared and the user can throw another dice.
'
' This program uses the PEEK command to read data from port B of
' the microcontroller.
'
'*****************************************************************
        symbol TRISB = 134           'Assign TRISB port RB
        symbol PORTB = 6             'Assign RB address
        symbol latch = 6             'Assign 6 to latch
        symbol secs3 = 3000          '3 secs=3000 ms
        symbol dice = B1             'Assign dice to B1
        symbol disp = B2             'Assign disp to B2


'
' Configure port RB pins.
'       Bits 0-6 are outputs
'       Bit 7 is input
'
        POKE TRISB,%10000000  'Set Port RB pins
        HIGH latch            'Set latch HIGH to start with
        POKE PORTB,0          'Clear display to start with
'
' Check switch status continuously.  The switch pin RB7 is
' normally held at logic HIGH with a pull-up resister.
' When the switch is pressed this pin goes to logic LOW.
'
start:  dice = 1                    'Set dice value to 1 to start with
loop:
        PEEK PORTB,B0               'Get port B data
        IF BIT7 = 0 THEN sw         'Goto sw if switch is pressed
        dice=dice+1                 'Increment the dice value
        IF dice = 7 then start      'If 7, reset the dice to 1
        GOTO loop                   'Go back and repeat
'
' We come here whenever the switch is pressed.
'
sw:
        GOSUB TIL311                'Go and display the dice value
        PAUSE secs3                 'Delay 3 seconds
        POKE PORTB,0               'Clear display
        GOTO loop                   'Go back to the beginning


'
' This subroutine displays the contents of variable dice on
' a TIL311 type hexadecimal alphanumeric display.
'
TIL311:
        disp = 64 + dice           'Set the latch
        POKE PORTB,disp            'Send data to the display
        LOW latch                  'Clear the latch
        HIGH latch                 'Set the latch
        RETURN                     'Return to main program
END
```

Fig. 5.25 Program listing of Project 9

PROJECT 10 – 7 Segment Display Driver

Function

This project shows how a 7 segment display can be interfaced to a microcontroller. In this project, a 7 segment display is connected to port B of the PIC microcontroller and a program is written to count up from 0 to 9 and display the data on the 7 segment display. The program runs in an endless loop and a 1 second delay is used between each output.

Circuit Diagram

Seven segment displays are used in many industrial and commercial applications. Basically the display consists of seven segments of LEDs, connected either as common anode or common cathode. In a common anode display the anodes of all the LED segments are connected together. Similarly, all the cathodes are connected together in a common cathode display. Segments in a 7 segment display are identified by giving them letters from a to g as shown in Fig. 5.26.

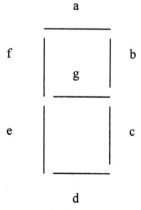

Fig. 5.26 Segments of a 7 segment display

Required characters are generated by turning on the appropriate LED segments. Table 5.3 shows the segments that should be turned on to generate the decimal numbers 0 to 9. A 1 in the table corresponds to the segment being on. In practical applications more than one display is used and the displays are multiplexed and driven together in order to minimize the power consumption and also simplify the design. Usually each display is scanned and refreshed for a short while and the scanning frequency is selected such that the overall display seems to be stationary.

The circuit diagram of Project 10 is shown in Fig. 5.27. A common anode type display is used in this project. The anode pins (3 and 8) are connected to a +5 V supply. Segments a to g are connected to port B of the microcontroller via 470 Ω

Table 5.3 Numbers displayed and segments

Number	g f e d c b a
0	0 1 1 1 1 1 1
1	0 0 0 0 1 1 0
2	1 0 1 1 0 1 1
3	1 0 0 1 1 1 1
4	1 1 0 0 1 1 0
5	1 1 0 1 1 0 1
6	1 1 1 1 1 0 0
7	0 0 0 0 1 1 1
8	1 1 1 1 1 1 1
9	1 1 0 0 1 1 1

current limiting resistors. Segment a is connected to bit 0 of port B (RB0), segment b to bit 1 of port B (RB1), segment c to bit 2 of port B (RB2) and so on.

Program Description

Table 5.4 shows the bit patterns (in hexadecimal notation) that should be sent to a 7 segment display in order to turn on the appropriate segments to display a particular

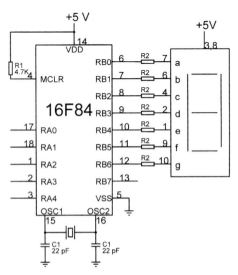

Fig. 5.27 Circuit diagram of Project 10

Table 5.4 Segments and corresponding bit patterns

Number	x g f e d c b a	Hex
0	0 0 1 1 1 1 1 1	3F
1	0 0 0 0 0 1 1 0	06
2	0 1 0 1 1 0 1 1	5B
3	0 1 0 0 1 1 1 1	4F
4	0 1 1 0 0 1 1 0	66
5	0 1 1 0 1 1 0 1	6D
6	0 1 1 1 1 1 0 0	7C
7	0 0 0 0 0 1 1 1	07
8	0 1 1 1 1 1 1 1	7F
9	0 1 1 0 0 1 1 1	67

number. Notice that in this table segment x is not used but included here so that the hexadecimal numbers can be derived easily as two 4-bit nibbles. For example, sending hexadecimal number 6D (or bit pattern 0110 1101) turns on the appropriate segments so that number 5 is displayed. Similarly, sending hexadecimal number 3F (bit pattern 0011 1111) displays number 0.

The following PDL describes the functions of the program:

Main Program

START

Configure port B pins as output

DO FOREVER

FOR count = 0 **TO** 9

Call subroutine SEVEN_SEGMENT to display count

Delay 1 second

NEXT

ENDDO

END

Subroutine SEVEN_SEGMENT

START

Find the 7 segment code corresponding to number to be displayed

Invert the code

Send the code to the display

Return

END

Program Listing

The full program listing is shown in Fig. 5.28. Variable *cnt* is assigned to location B0 and variable *disp* is a temporary variable, assigned to location B1. Port B pins are configured as outputs. A *FOR* loop is then formed where variable *cnt* is changed from 0 to 9 and subroutine SEVEN_SEGMENT is called to display the value of *cnt*. After a 1 second delay the program loop is repeated. The following data is displayed by the program:

0 1 2 3 4 5 6 7 8 9 0 1 . . .

Subroutine SEVEN_SEGMENT displays the number in variable *cnt* on the 7 segment display. The hexadecimal bit patterns corresponding to each decimal number (see Table 5.4) are stored in a look-up table and command LOOKUP is used to index this table and extract the correct bit patterns. For example, if *cnt* is 0, variable *disp* is loaded with the hexadecimal number 3F. If *cnt* is 1, variable *disp* is loaded with the hexadecimal number 06 and so on. The display used in this example is a common anode type where current is sourced. Thus, a segment LED is turned on when the corresponding port B output pin is at logic LOW. As a result of this, it is necessary to invert the data before sending it to the display. This is done by exclusive-OR'ing the data with the hexadecimal number FF (i.e. all 1s). As shown below, exclusive-OR'ing a bit with 1 always inverts that bit (in the example given below, the "^" sign denotes the exclusive-OR operation):

0 ^ 1 = 1

1 ^ 1 = 0

PROJECT 11 – Light Dimming with PWM

Function

This project shows how we can send out pulse width modulated data (PWM) on a specified pin of the microcontroller. In PWM the output is cycled between HIGH

```
'*****************************************************************
'
'           PROJECT:           PROJECT10
'           FILE:              PROJ10.BAS
'           DATE:              August 2000
'           PROCESSOR:         PIC16F84
'           COMPILER:          PIC BASIC
'
'
' This is a 7 segment display interface project.
' The display is connected to port B of the PIC
' microcontroller and counts up from 0 to 9 with
' 1 second delay between each count.
'
'*****************************************************************
symbol TRISB = 134              'Assign TRISB port RB
symbol PORTB = 6                'Assign RB address
symbol second = 1000            '1 second=1000 ms
symbol cnt = B0                 'Assign B0 to variable cnt
symbol disp = B1                'Assign B1 to variable disp

'
' Configure port B pins
'
          POKE TRISB,0          'Configure port B as output

' Start of main program
'
loop:
          FOR cnt = 0 TO 9                'Do 0 to 9
                GOSUB SEVEN_SEGMENT       'Go and display cnt
                PAUSE second              'Delay a second
          NEXT cnt                        'End of for loop

          GOTO loop                       'Go and do forever


'
' This subroutine displays the number in variable cnt
' on a 7-segment display.  The display is common anode
' type where current is sourced and thus, the data is
' inverted before it is sent to the display.
'
SEVEN_SEGMENT:
          LOOKUP cnt,($3F,$06,$5B,$4F,$66,$6D,$7C,$07,$7F,$67),disp
          disp=disp ^ $FF        'Invert the data
          POKE PORTB,disp        'Send the data to the display
          RETURN                 'Return to main program

    END
```

Fig. 5.28 Program listing of Project 10

and LOW over a short period of time. This kind of output can be approximated to an analogue output data. The longer the output is HIGH, the greater the average analogue output is. Figure 5.29 shows a typical PWM signal. The percentage of time that the signal is HIGH over during its period is known as the duty cycle of the signal. The signal in Fig. 5.29 has a duty cycle of 25%. The average output voltage of a PWM signal can be approximated to its peak voltage times the duty cycle. For example, the signal in Fig. 5.29 has an average value of $5\,V \times 0.25 = 1.25\,V$. By varying the duty cycle we can vary the average value of the signal.

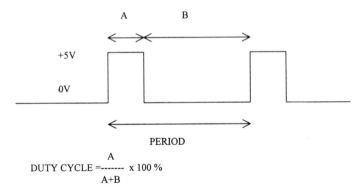

Fig. 5.29

In this project, an LED is connected to the microcontroller. PWM signal is then sent to the LED with varying duty cycle and as the average voltage at the LED changes, the LED dims slowly.

Circuit Diagram

The circuit diagram of this project is the same as in Project 1 (i.e. Fig. 5.1).

Program Description

The program sends a PWM signal to the LED with varying duty cycle. All the possible duty cycles (0 to 255) are sent to the LED. The following PDL describes the operation of the program:

START

      DO for all possible duty cycles

          Send a PWM signal to the LED

      ENDDO

   END

Program Listing

The listing of Project 11 is shown in Fig. 5.30. A FOR loop is formed with the duty cycle changing from 0 to 255. PIC BASIC command PWM is used to send a PWM signal to the LED. The cycle time of the signal is set to 2 ms. When power is applied to the microcontroller, the LED is fully on. The LED is then gradually dimmed and then turned off after a short while.

Figure 5.31 shows a modification of the basic program where two FOR loops are used. The LED is initially dimmed and then brightened again. This is performed in a continuous loop.

```
'*****************************************************************
'
'          PROJECT:          PROJECT11
'          FILE:             PROJ11.BAS
'          DATE:             August 2000
'          PROCESSOR:        PIC16F84
'          COMPILER:         PIC BASIC
'
'
' This is a simple LED project using the PWM command.  The
' intensity of an LED light connected to port RB0 is first
' bright and then dimmed and then set to be bright and so on.
' This process is repeated continuously by varying the duty
' cycle of the PWM signal output from the microcontroller.
'
'*****************************************************************
'
symbol  led = 0                        'Assign led to 0 (RB0)
symbol  duty = B0                      'Assign duty to location B0

        OUTPUT led                     'Set pin 0 as output
loop:
        FOR duty = 0 TO 255            'Start of loop
           PWM led,duty,2              'Send PWM signal
        NEXT duty                      '
        GOTO loop                      'Go back and repeat

    END
```

Fig. 5.30

```
'*****************************************************************
'
'          PROJECT:          PROJECT11
'          FILE:             PROJ11-1.BAS
'          DATE:             August 2000
'          PROCESSOR:        PIC16F84
'          COMPILER:         PIC BASIC
'
'
' This is a simple LED project using the PWM command.  The
' intensity of an LED light connected to port RB0 is first
' bright and then dimmed gradually until it is off.  The
' light intensity is then increased gradually until it is
' fully on, and so on.
' This process is repeated continuously by varying the duty
' cycle of the PWM signal output from the microcontroller.
'
'*****************************************************************
'
symbol  led = 0                        'Assign led to 0 (RB0)
symbol  duty = B0                      'Assign duty to location B0

        OUTPUT led                     'Set pin 0 as output
loop:
        FOR duty = 0 TO 255            'Start of loop
           PWM led,duty,2              'Send PWM signal
        NEXT duty                      '

        FOR duty = 255 to 0 step -1    'Start of loop
           PWM led,duty,2              'Send PWM signal
        NEXT duty                      '

        GOTO loop                      'Go back and repeat

    END
```

Fig. 5.31

PROJECT 12 – Liquid Crystal Display (LCD) Interface

One thing all microcontrollers lack is some kind of video display. A video display would make a microcontroller much more user friendly as it will enable text messages and numeric values to be output in a more versatile manner than the 7 segment displays, LEDs, or alphanumeric displays. Standard video displays require complex interfaces and their cost is relatively high. LCDs are alphanumeric displays which are frequently used in microcontroller-based applications. These display devices come in different shapes and sizes. Some LCDs have 40 or more character lengths with the capability to display several lines. Some other LCD displays can be programmed to display graphics images. Some modules offer colour displays while some others incorporate back lighting so that they can be viewed in dimly lit conditions.

There are basically two types of LCDs as far as the interface technique is concerned: parallel LCDs and serial LCDs. Parallel LCDs (e.g. Hitachi HD44780) are connected to the microcontroller circuitry such that the data is transferred to the LCD unit using more than one data line, and four or eight data lines are very common. Serial LCDs are connected to the microcontroller using only one data line and data is transferred to the LCD using the standard RS-232 asynchronous data communication protocols. Serial LCDs are much easier to use but they usually cost more than the parallel ones.

The programming of a parallel LCD is usually a complex task and requires a good understanding of the internal operation of the LCDs, including the timing diagrams. Fortunately, PIC BASIC PRO language provides special commands for displaying data on the HD44780 type parallel LCDs. All the user has to do is connect the LCD to the appropriate I/O ports and then use these special commands to simply send data to the LCD. The standard PIC BASIC language does not provide any special commands for programming the parallel LCDs and the programming of these LCDs using the PIC BASIC language is beyond the scope of this book. The programming of serial LCDs using the PIC BASIC language is covered in Chapter 8 in detail.

In the remainder of this chapter we shall be looking at the characteristics of the HD44780 (or compatible) parallel LCD modules and also describe several projects using these LCDs. First, let's look at the characteristics of the parallel LCDs and how they can be programmed using the PIC BASIC PRO language.

HD44780 LCD Module

HD44780 is one of the most popular LCD modules used in industry and also by hobbyists. This module is monochrome and comes in different shapes and sizes. Modules with line lengths of 8, 16, 20, 24, 32, and 40 characters can be selected. Depending upon the model chosen, the display width can be selected as 1, 2, or 4 lines. The display provides a 14-pin connector to interface to a microcontroller. Table 5.5 shows the pinout and pin functions for the LCD. Below is a summary of the pin functions.

Table 5.5 Pinout of the HD44780 LCD module

Pin No	Name	Function
1	V_{SS}	Ground
2	V_{DD}	+ ve supply
3	V_{EE}	Contrast
4	RS	Register select
5	R/W	Read/write
6	E	Enable
7	D0	Data bit 0
8	D1	Data bit 1
9	D2	Data bit 2
10	D3	Data bit 3
11	D4	Data bit 4
12	D5	Data bit 5
13	D6	Data bit 6
14	D7	Data bit 7

V_{SS} is the 0 V supply or ground. The V_{DD} pin should be connected to the positive supply. Although the manufacturers specify a 5 V d.c. supply, the modules will usually work with as low as 3 V or as high as 6 V.

Pin 3 is named as V_{EE} and this is the contrast control pin. This pin is used to adjust the contrast of the device and it should be connected to a variable voltage supply. A potentiometer is usually connected between the power supply lines with its wiper arm connected to this pin so that the contrast can be adjusted. This pin can be connected to ground for most applications.

Pin 4 is the Register Select (RS) and when this pin is LOW, data transferred to the display is treated as commands. When RS is HIGH, character data can be transferred to and from the module.

Pin 5 is the Read/Write (R/W) line. This pin is pulled LOW in order to write commands or character data to the LCD module. When this pin is HIGH, character data or status information can be read from the module.

Pin 6 is the Enable (E) pin which is used to initiate the transfer of commands or data between the module and the microcontroller. When writing to the display, data is transferred only on the HIGH to LOW transition of this line. When reading from the display, data becomes available after the LOW to HIGH transition of the enable pin and this data remains valid as long as the enable pin is at HIGH.

Pins 7 to 14 are the eight data bus lines (D0 to D7). Data can be transferred between the microcontroller and the LCD unit using either a single 8-bit byte, or as two 4-bit nibbles. In the latter case, only the upper four data lines (D4 to D7) are used. 4-bit mode has the advantage that fewer I/O lines are required to communicate with the LCD.

Connecting the LCD to the Microcontroller

PIC BASIC PRO compiler assumes that the LCD is connected to specific pins of the microcontroller unless told otherwise. It assumes that the LCD will be used with a 4-bit bus with data lines D4–D7 connected to microcontroller PORTA.0–PORTA.3, Register Select line to PORTA.4, and Enable pin to PORTB.3. A two-line display is also assumed. When these connections are made between the microcontroller and the LCD, we can simply use the LCDOUT command to send data to the LCD module. Note that the connection between the microcontroller and the LCD can be changed by using a set of DEFINE commands to assign the LCD pins to the PIC microcontroller (see PIC BASIC PRO Compiler manual for more information).

Programming the LCD

The LCDOUT command displays items (text or numbers) on the LCD display. In addition to the normal data display, a number of LCD display functions can be performed as shown in Table 5.6. When using these display functions, the data sent to the LCD should be preceded with the hexadecimal command $FE. For example, the command:

LCDOUT $FE, $C0

moves the cursor to the beginning of the second line. Similarly, the command:

LCDOUT $FE, 1

Table 5.6 LCD commands

Command	Operation
$FE, 1	Clear display
$FE, 2	Return home
$FE,$0C	Cursor off
$FE,$0E	Underline cursor on
$FE,$0F	Blinking cursor on
$FE,$10	Move cursor left one position
$FE,$14	Move cursor right one position
$FE,$C0	Move cursor to beginning of second line
$FE,$94	Move cursor to beginning of third line
$FE,$D4	Move cursor to beginning of fourth line

clears the display. If a hash sign (#) precedes an item, the ASCII representation for each digit is sent to the LCD. It is important to realize that a program should wait for at least half a second before sending the first command to an LCD. This is because it can take quite a while before the LCD initializes itself.

Table 5.7 shows the LCD character table. Using the codes in this table we can send non-ASCII characters and different signs to the LCD.

Table 5.7 LCD character table

Upper 4 bits / Lower 4 bits	0 0000	1 0001	2 0010	3 0011	4 0100	5 0101	6 0110	7 0111	8 1000	9 1001	A 1010	B 1011	C 1100	D 1101	E 1110	F 1111
0 0000	CG RAM (1)			0	@	P	`	p				ー	タ	ミ	α	p
1 0001	CG RAM (2)		!	1	A	Q	a	q			。	ア	チ	ム	ä	q
2 0010	CG RAM (3)		"	2	B	R	b	r			「	イ	ツ	メ	β	θ
3 0011	CG RAM (4)		#	3	C	S	c	s			」	ウ	テ	モ	ε	∞
4 0100	CG RAM (5)		$	4	D	T	d	t			、	エ	ト	ヤ	μ	Ω
5 0101	CG RAM (6)		%	5	E	U	e	u			・	オ	ナ	ユ	σ	ü
6 0110	CG RAM (7)		&	6	F	V	f	v			ヲ	カ	ニ	ヨ	ρ	Σ
7 0111	CG RAM (8)		'	7	G	W	g	w			ァ	キ	ヌ	ラ	g	π
8 1000	CG RAM (1)		(	8	H	X	h	x			ィ	ク	ネ	リ	√	x̄
9 1001	CG RAM (2)		)	9	I	Y	i	y			ゥ	ケ	ノ	ル	⁻¹	y
A 1010	CG RAM (3)		*	:	J	Z	j	z			ェ	コ	ハ	レ	j	千
B 1011	CG RAM (4)		+	;	K	[	k	{			ォ	サ	ヒ	ロ	ˣ	万
C 1100	CG RAM (5)		,	<	L	¥	l	|			ャ	シ	フ	ワ	¢	円
D 1101	CG RAM (6)		-	=	M	]	m	}			ュ	ス	ヘ	ン	£	÷
E 1110	CG RAM (7)		.	>	N	^	n	→			ョ	セ	ホ	゛	ñ	
F 1111	CG RAM (8)		/	?	O	_	o	←			ッ	ソ	マ	゜	ö	█

In the remainder of this chapter we shall be looking at a number of LCD-based projects where an LCD is used to display text as well as numeric data.

Function

This project shows how we can connect an LCD to a PIC microcontroller and then display a simple text message. The message displayed in this project is "PIC LCD – ROW 1" on the first row, and "PIC LCD – ROW2" on the second row of the LCD.

Circuit Diagram

Figure 5.32 shows how an LCD can be connected to a PIC microcontroller. The following standard connections are made between the LCD module and the microcontroller:

LCD	Microcontroller
D4	RA0
D5	RA1
D6	RA2
D7	RA3
E	RB3
RS	RA4

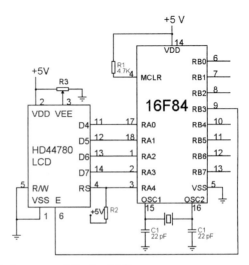

Fig. 5.32 Circuit diagram of Project 12

A 20K potentiometer is used to adjust the contrast of the LCD. Also note that a 10K pull-up resistor is used at pin RA4 of the microcontroller since this is an open-drain pin.

Program Description

The program is very simple as most of the LCD functions are included in the PIC BASIC PRO language. The following PDL describes the operation of the program:

START

> Wait a second until LCD initializes

> Clear the LCD display

> Display message "PIC LCD – ROW1"

> Move to second line

> Display message "PIC LCD – ROW2"

END

Program Listing

The full program listing is shown in Fig. 5.33. After the initialization of the LCD, the LCD display is cleared using the command:

LCDOUT $FE, 1

```
'********************************************************************
'
'          PROJECT:          PROJECT12
'          FILE:             PROJ12.BAS
'          DATE:             August 2000
'          PROCESSOR:        PIC16F84
'          COMPILER:         PIC BASIC PRO
'
'
' This is a simple LCD display project.  A HD44780 type parallel
' LCD is connected to a PIC microcontroller.  The LCD chosen is
' a 2 row type (model LM016L).  The program simply writes the
' following message to the display:
'
'          PIC LCD - ROW1
'          PIC LCD - ROW2
'
'********************************************************************
symbol second = 1000

          PAUSE second                 'Wait until the LCD initializes
          LCDOUT $FE,1                 'Clear the LCD
          LCDOUT "PIC LCD - ROW1"      'Display message
          LCDOUT $FE,$C0               'Move to second row
          LCDOUT "PIC LCD - ROW2"      'Display message
          STOP                         'End of program
     END
```

Fig. 5.33 Program listing of Project 12

The text message "PIC LCD – ROW1" is then displayed on the first row. The command:

    LCDOUT $FE, $C0

moves the cursor to the beginning of the second row where the message "PIC LCD – ROW2" is displayed.

Components Required

In addition to the components required by the basic microcontroller circuit, the following components will be required for Project 11:

    R2   10K, 0.125 W resistor
    R3   20K potentiometer
          HD44780 LCD

PROJECT 13 – LCD Counter

Function

This project shows how we can set up a simple counter and display the count on an LCD display. Data is displayed in the following format:

    COUNT: nn

where nn is incremented every second from 00 to 100.

Circuit Diagram

The circuit diagram of this project is the same as in Fig. 5.32 where the LCD is connected to PORT A of the microcontroller.

Program Description

The program counts from 0 to 100 with a 1 second delay between each count and the result is displayed on the LCD. The following PDL describes the operation of the program:

START

> Initialize the LCD
>
> Clear the LCD display
>
> **DO** 100 times
>
>> Display 'COUNT : current count'
>>
>> Return the cursor to home position
>>
>> Delay 1 second
>
> **ENDDO**
>
> **END**

Program Listing

The full program listing is shown in Fig. 5.34. A *FOR* loop is set up to count from 1 to 100 with variable *cnt* and the value of *cnt* is displayed on the LCD. A 1 second delay is inserted between each count using the *PAUSE second* command.

```
'*****************************************************************
'
'          PROJECT:          PROJECT13
'          FILE:             PROJ13.BAS
'          DATE:             August 2000
'          PROCESSOR:        PIC16F84
'          COMPILER:         PIC BASIC PRO
'
'
' This is a simple LCD display project.  A HD44780 type parallel
' LCD is connected to a PIC microcontroller.  The LCD chosen is
' a 2 row type (model LM016L).  The program counts from 1 to 100
' and the count is displayed on the LCD.  The data is displayed
' in the following format with 1 second delay between each count:
'
'          COUNT: nn
'
' where nn = 00 to 100
'
'*****************************************************************
symbol second = 1000                    'seconds delay
cnt var byte                            'count

        PAUSE second                    'Wait until the LCD initializes
        LCDOUT $FE,1                    'Clear the LCD
        cnt=0                           'Clear cnt
        FOR cnt=1 TO 100                'Start of loop
           LCDOUT "COUNT: ",#cnt        'Display text and count
           LCDOUT $FE,2                 'Return to first line (home)
           PAUSE second                 'Delay a second
        NEXT cnt                        'End of loop

        STOP                            'End of program
END
```

Fig. 5.34 Program listing of Project 13

PROJECT 14 – Elapsed Seconds Counter with LCD Display

Function

This project shows how we can design a simple elapsed seconds counter and display the accumulated seconds on an LCD. Three push-button switches named START, STOP, and RESET are connected to port B of the microcontroller. When START is pressed, counting starts and the count is updated every second on the LCD display. When STOP is pressed, counting stops. Pressing the START button at this point resumes the count. When RESET is pressed, the LCD is cleared and the process repeats, i.e. the program waits for the user to press the START button.

Circuit Diagram

As shown in Fig. 5.35, the LCD is connected to port A of the microcontroller. The START button is connected to pin 0 of port B (RB0), the STOP button is connected to pin 1 of port B (RB1), and finally, the RESET button is connected to pin 2 of port B (RB2). Pull-up resistors are used at all three switch inputs to hold the port pins at logic HIGH.

Program Description

The program should initialize the LCD and then wait until the user presses the START button. A count should then be set up with a rate of 1 second and the result

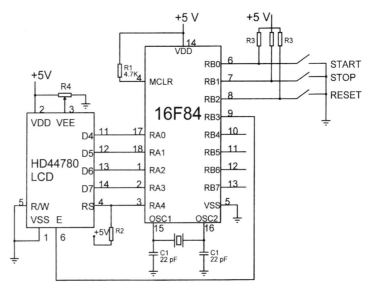

Fig. 5.35 Circuit diagram of Project 14

should be displayed on the LCD. When the STOP button is pressed, the counting should stop. Pressing the RESET button should clear the display and jump to the point in the program where counting can be restarted. The following PDL describes the operation of the program:

START

    Initialize the LCD

    Display "READY" message

    Initialize count

    Clear start_flag

    DO FOREVER

        IF START button pressed **THEN**

            Set start_flag

        ENDIF

        IF STOP button is pressed **THEN**

            Clear start_flag

        ENDIF

        IF RESET button is pressed **THEN**

            Clear start_flag

            Clear LCD

            Display "READY" message

        ENDIF

        IF start_flag is set **THEN**

            Increment count

            Display count on LCD

            Delay 1 second

        ENDIF

    ENDDO

    END

Program Listing

The full program listing is shown in Fig. 5.36. Push-buttons START, STOP, and RESET are assigned to port B pins using the var statements. The LCD is then initialized and the message "READY" is displayed, waiting for the user to start the counting. BUTTON commands are used to test the status of the push-button switches. When the START switch is pressed, the program jumps to label *lbl_start*. When the STOP switch is pressed, the program jumps to label *lbl_stop*, and when the RESET switch is pressed, the program continues at label *lbl_reset*.

Counting continues as long as variable *start_flag* is set. At label *lbl_start*, counter variable *cnt* is incremented and its value is displayed on the LCD in the following format:

SECONDS = nnn

After a 1 second delay, the program jumps back to recheck the status of the push-button switches. At label *lbl_stop*, variable *start_flag* is cleared to stop the counting and the program jumps back to recheck the status of the push-button switches.

At label *lbl_reset*, the *start_flag* is cleared, the LDC display is cleared and the program jumps back to recheck the switches.

```
'*****************************************************************
'
'        PROJECT:        PROJECT14
'        FILE:           PROJ14.BAS
'        DATE:           August 2000
'        PROCESSOR:      PIC16F84
'        COMPILER:       PIC BASIC PRO
'
'
' This is a simple LCD display project.  A HD44780 type parallel
' LCD is connected to a PIC microcontroller.  The LCD chosen is
' a 2 row type (model LM016L).  When the program starts, the
' LCD displays the message "READY".  Three push-button switches,
' named START, STOP, and RESET are connected to RB0, RB1, and RB2
' respectively.  When START is pressed, program starts counting
' with 1 second delay between each count and the count is shown
' on the display.  When STOP is pressed, the counting stops.  The
' counting can continue if START is pressed again.  When RESET is
' pressed the counting stops, LCD is cleared and the READY message
' is displayed on the LCD.
'
'*****************************************************************

symbol second = 1000             'seconds delay
strt var PORTB.0                 'START button
stp var PORTB.1                  'STOP button
rst var PORTB.2                  'RESET button
cnt var word                     'count
start_flag var byte              'start flag
tmp1 var byte                    'temporary variable
tmp2 var byte                    'temporary variable
tmp3 var byte                    'temporary variable

        PAUSE second             'Wait until the LCD initializes
        LCDOUT $FE,1             'Clear the LCD
        LCDOUT "READY"          'Display READY message

        TRISB=%00000111          'Set RB0,RB1,RB2 as inputs
        cnt=0                    'Clear count
        start_flag=0             'Clear start count
```

Fig. 5.36 Program listing of Project 14

Components Required

In addition to the components required by the basic microcontroller circuit, the following components will be required for Project 14:

R2   10 K, 0.125 W resistor

R3   10 K, 0.125 W resistors (3 off)

R4   20 K potentiometer

HD44780 LCD

```
loop:
            tmp1=0                              'Clear temp1,2,3 for BUTTON
            tmp2=0                              '
            tmp3=0                              '
            BUTTON strt,0,255,0,tmp1,1,lbl_start
            BUTTON stp,0,255,0,tmp2,1,lbl_stop
            BUTTON rst,0,255,0,tmp3,1,lbl_reset

            IF start_flag = 1 THEN
                    GOTO lbl_start
            ENDIF

            GOTO loop                           'Go back and repeat

'
' We jump here when the START button is pressed.  A new count
' is obtained and the current seconds is displayed on the LCD.
'
lbl_start:
            start_flag=1                        'Set start flag to 1
            LCDOUT $FE,2                         'Return to home position
            LCDOUT "SECONDS = ",#cnt            'Display on LCD
            cnt=cnt+1                           'Increment count
            PAUSE second                        'Delay a second
            GOTO loop                           'Go back and repeat

'
' We jump here when the STOP button is pressed.  start flag is
' cleared and program jumps back to the loop.  Counting can
' resume if the START button is pressed again.
'
lbl_stop:
            start_flag=0                        'Clear start flag
            GOTO loop                           'Go back and repeat

'
' We jump here when the RESET button is pressed. Counting stops
' start flag is cleared, and the READY message is displayed on
' the LCD.
'
lbl_reset:
            cnt=0                               'Clear count
            start_flag=0                        'Clear start flag
            LCDOUT $FE,1                         'Clear LCD display
            LCDOUT "READY"                       'Display READY message
            GOTO loop                           'Go back and repeat

            END
```

Fig. 5.36   (Continued)

PROJECT 15 – Light Sensor with Analogue-to-Digital Converter and LCD Output

Function

This project demonstrates how we can use the analogue-to-digital (A/D) converter port of a PIC16C71 type microcontroller. Port RA0 is configured as an analogue port and a light sensor is connected to this port. The light sensor produces a voltage proportional to the light intensity. This voltage is converted into digital form and displayed on an LCD.

Circuit Diagram

The circuit diagram of Project 15 is shown in Fig. 5.37. Since we need to use port A for analogue input, the LCD is connected to port B of the microcontroller as follows:

Microcontroller	LCD
RB4	D4
RB5	D5
RB6	D6
RB7	D7
RB1	RS
RB0	E

As a result of this non-standard connection between the microcontroller and the LCD, we have to define this connection at the beginning of our program.

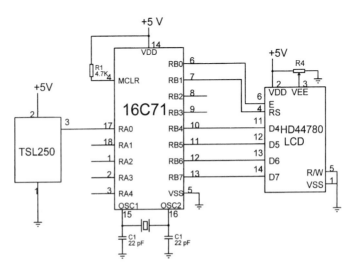

Fig. 5.37 Circuit diagram of Project 15

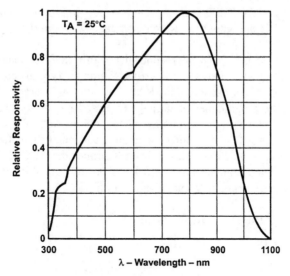

Fig. 5.38 Spectral responsivity of TSL250

A TSL250 type light-to-voltage sensor is connected to pin 0 of port A (RA0). This is a 3-pin optical sensor IC which combines a photodiode and an amplifier on a single monolithic IC. The output voltage is directly proportional to the light intensity on the sensor. The device can be operated from a supply voltage as low as 3 V. Figure 5.38 shows the spectral responsivity of this sensor. The sensor is most responsive around the wavelength of 800 nm. The output voltage characteristic of the sensor is shown in Fig. 5.39 as a function of the irradiance.

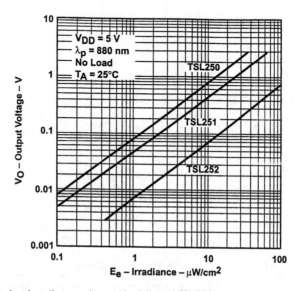

Fig. 5.39 Output voltage characteristic of TSL250

Pin 1 of the TSL250 is connected to ground, pin 2 is connected to the +5 V supply and pin 3 is the output pin which is connected to pin 0 of port A.

PIC16C71 is an 18-pin microcontroller with four 8-bit analogue-digital-converter (A/D) channels, and four interrupt sources. Typical conversion time is 20 microseconds per channel. Pins RA0–RA3 of 16C71 are analogue channels which can also be programmed as digital I/O pins. Two control registers, ADCON0 and ADCON1, are used to program the A/D functions. ADCON1 defines the various bits to select the A/D clock source, analogue channels, to start the conversion, and the conversion completion status. Table 5.8 defines the ADCON0 bits. The A/D clock can be derived from the system clock or from an internal RC oscillator. ADCON bits 6 and 7 select the clock source as follows:

ADCON0_7	ADCON0_6	
0	0	$f_{OSC/2}$
0	1	$f_{OSC/8}$
1	0	$f_{OSC/32}$
1	1	f_{RC}

A/D channels are selected using bits 4 and 5 of the ADCON0 register:

ADCON_4	ADCON_3	
0	0	Channel 0
0	1	Channel 1
1	0	Channel 2
1	1	Channel 3

Table 5.8 ADCON0 bit definitions

ADCON_7	A/D clock select
ADCON_6	A/D clock select
ADCON_5	Read/write (general purpose)
ADCON_4	A/D channel select
ADCON_3	A/D channel select
ADCON_2	Set to 1 to start conversion
ADCON_1	A/D conversion complete
ADCON_0	A/D converter is shut off and consumes no current (=0)

Register ADCON1 is two bits wide and these bits configure the function of pins RA0–RA3:

ADCON1_1	ADCON_0	
0	0	RA0–RA3 analogue inputs
0	1	RA0–RA2 analogue inputs
		RA3 reference input
1	0	RA0–RA1 analogue inputs
		RA2–RA3 digital I/O
1	1	RA0–RA3 digital I/O

Program Description

At the beginning of the program, microcontroller port B pins should be assigned to the LCD. This is necessary since the LCD is connected to port B and not in the standard configuration. A/D clock source and the A/D width should then be defined. Variable data should then be read from the A/D channel and displayed on the LCD with a 1 second delay between each sample. The following PDL describes the operation of the program:

START

    Define the microcontroller–LCD interface

    Initialize the A/D

DO FOREVER

    Get A/D data from the light sensor

    Display data on LCD

ENDDO

END

Program Listing

The complete program listing is shown in Fig. 5.40. The interface between the microcontroller and the LCD is described using a set of DEFINE statements at the beginning of the program:

DEFINE LCD_DREG PORTB: Define that port B is used to connect to the data lines of LCD.

DEFINE LCD_DBIT 4:   Define that the upper 4 bits of the port are used to connect to the LCD data lines.

DEFINE LCD_RSREG PORTB: Define that port B is used to control the RS pin
                           of the LCD.

DEFINE LCD_RSBIT 1:   Define that bit 1 of port B is used to control the
                           RS pin of the LCD.

DEFINE LCD_EREG PORTB: Define that port B is used to control the Enable
                           pin of the LCD.

DEFINE LCD_EBIT 0:   Define that pin 0 of port B is used to control the
                           Enable pin of the LCD.

```
'*********************************************************************
'
'          PROJECT:        PROJECT15
'          FILE:           PROJ15.BAS
'          DATE:           August 2000
'          PROCESSOR:      PIC16C71
'          COMPILER:       PIC BASIC PRO
'
'
' This is a simple LCD based light sensor project. A HD44780
' type parallel LCD is connected to a PIC microcontroller.
' Also, a TLS250 type light-to-voltage converter sensor is
' connected to one of the analog ports of the PIC microcontroller.
' The program converts the analog output of the light sensor to
' digital and then displays the data on the LCD display.  The
' data displayed is proportional to the light intensity and needs
' to be calibrated for an accurate measurement.
'
'*********************************************************************
'
' Assign LCD pins to PORT B
'
DEFINE LCD_DREG PORTB                   'Using PORT B for LCD data
DEFINE LCD_DBIT 4                       'Using upper 4 bits
DEFINE LCD_RSREG PORTB                  'Using PORT B for RS reg
DEFINE LCD_RSBIT 1                      'Using PORTB.1 for RS pin
DEFINE LCD_EREG PORTB                   'Using PORT B for E reg
DEFINE LCD_EBIT 0                       'Using PORTB.0 for E pin
DEFINE LCD_BITS 4                       'LCD is in 4 bit data mode
DEFINE LCD_LINES 2                      'LCD has 2 lines
'
' Define A/D converter pins
'
DEFINE ADC_BITS 8                       'Using 8 bit conversion
DEFINE ADC_CLOCK 3                      'Clock source=3 (RC)

new_data var byte                       'A/D converter data

symbol second = 1000                    'seconds delay

        PAUSE second                    'Wait until the LCD initializes
        LCDOUT $FE,1                    'Clear the LCD

        TRISA = %11111111              'Configure all pins as inputs
        ADCON1 = 0                      'Set PortA 0-3 to analog
loop:
        ADCIN 0,new_data                'Read analog channel 0 to B0
        LCDOUT $FE,1                    'Clear LCD
        LCDOUT "Light = ",#new_data     'Display new data on LCD
        PAUSE second                    'Delay a second
        GOTO loop                       'Go back and repeat

    END
```

Fig. 5.40 Program listing of Project 15

DEFINE LCD_BITS 4:   Define that the LCD is in 4-bit data mode.

DEFINE LCD_LINES 2:   Define that the LCD has two lines.

The A/D converter width and the clock source are then defined using the following statements at the beginning of the program:

DEFINE ADC_BITS 8:   Define that we are using 8-bit conversion.

DEFINE ADC_CLOCK 3:   Define that the A/D clock source is
RC (ADCON0_7 = 1, ADCON0_6 = 1)

Variable *new_data* is then declared and this variable is used to store the converted digital data. TRISA is used to configure the analogue channels as inputs and all the RA0–RA3 inputs of the microcontroller are assigned as analogue inputs (using ADCON1). Data is received and displayed on the LCD using the following commands:

ADCIN 0,new_data

LCDOUT "Light =",#new_data

Components Required

In addition to the components required by the basic microcontroller circuit, the following components will be required for Project 15:

TSL250 light-to-voltage sensor

HD44780 LCD

R4   20K potentiometer

PROJECT 16 – Elapsed Time Counter with Timer Interrupt

Function

This project demonstrates how we can use the interrupt service routine and the internal timer of the PIC microcontrollers to design an elapsed time counter. Two push-button switches, named START and STOP are connected to the micro-controller. In addition, an LCD display is connected to show the elapsed time. When START is pressed, the timer is loaded and the program generates a seconds count based upon the timer overflow. A timer interrupt service routine is used to calculate the elapsed time. When STOP is pressed the count is stopped.

Circuit Diagram

The circuit diagram of Project 16 is shown in Fig. 5.41. The microcontroller is connected to the LCD as in Project 15 (Fig. 5.37). Push-button switches START and STOP are connected to pins 2 and 3 of port B respectively using pull-up resistors.

Timer 0 (TMR0) is an 8-bit timer which counts up and it can be programmed to generate an interrupt when it changes from 255 to 0. The timer is controlled by two registers, OPTION_REG and INTCON. OPTION_REG is a readable and writable register which contains various control bits to configure the timer, prescaler, external interrupt, and port B pull-ups. Bits 0–3 of the OPTION_REG control the prescaler assignment which is used to derive the clock for the timer. Bit 3 of this register should be 0 to assign the prescaler to the timer. Bits 0–2 define the timer clock rate as follows:

Bit value	Timer rate
000	1:2
001	1:4
010	1:8
011	1:16
100	1:32
101	1:64
110	1:128
111	1:256

Register INTCON contains various enable bits for the interrupt sources. Bit 7 is the global interrupt enable bit and this bit should be set to 1 to enable interrupts on a global basis. Bit 5 is the timer interrupt enable bit and this bit should be set to 1 to enable timer interrupts. Bit 2 is set by the microcontroller whenever the timer generates an interrupt. This bit should be cleared to 0 by software before a timer interrupt can be accepted by the microcontroller.

Program Description

The program should define the connection between the microcontroller and the LCD since the LCD is not connected in a standard way. The START button should then be scanned continuously. When this button is pressed, the program should load the timer with an appropriate value. When the timer overflows a timer interrupt will be generated and the program will jump the interrupt service routine (ISR). Inside the ISR, the program should check if a second has elapsed since the last update and

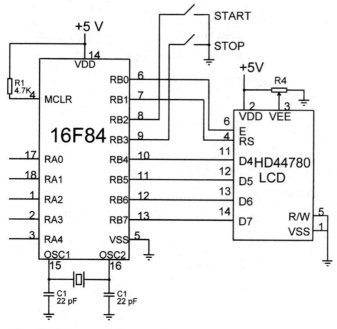

Fig. 5.41 Circuit diagram of Project 16

if so it should update the seconds, minutes, and the hours. The main program should update the LCD every second.

The following PDL describes the operation of the program:

Main Program

> **START**
>
>> Define the microcontroller–LCD interface
>>
>> Declare the interrupt service routine
>>
>> **DO FOREVER**
>>
>>> **IF START** button is pressed **THEN**
>>>
>>>> Load timer
>>>>
>>>> Enable interrupts
>>>
>>> **END IF**
>>>
>>> **IF** update flag is set **THEN**
>>>
>>>> Update LCD display
>>>
>>> **ENDIF**

          IF STOP button is pressed **THEN**

             Stop timer

             Disable timer interrupt

        ENDIF

      ENDDO

    END

Interrupt Service Routine

    START

        IF a seond has elapsed since last time **THEN**

             Set update flag

             Update seconds

             Update minutes

             Update hours

        ENDIF

    END

Program Listing

The complete program is shown in Fig. 5.42. The interface between the microcontroller and the LCD are described using a set of DEFINE statements as in Project 15. Variables *hour*, *minute*, and *second* are declared as bytes. The TRISB command is used to configure bits 2 and 3 of port B as inputs. When the start button is pressed (PORTB.2 = 0), the INTCON register is loaded with hexadecimal value $A0 (bit pattern 10100000) and the timer register (TMR0) is loaded with 131. Loading INTCON sets bits 5 and 7 which enables the timer interrupt as well as the global interrupts. The OPTION_REG register is loaded with 2 to select a prescaler value of 8. With a 4 MHZ clock, the basic instruction timing is 1 microsecond. Using a prescaler value of 8 we have the timer clock rate of 8 microseconds. Thus, if we load the timer with 125 then interrupts will be generated at every 125 × 8 = 1000 microseconds, i.e. every millisecond. Since the timer counts up to 255 and then resets back to 0, the value to be loaded into the timer register is 256 − 125 = 131 (i.e. a count of 131 before an interrupt is generated).

The Interrupt Service Routine (ISR) is entered at every millisecond. Variable *cnt* counts the number of entries and when *cnt* = *1000* (i.e. a second) then variables *second*, *minute*, and *hour* are updated accordingly. Variable *update* is also set to 1 so that the main program updates the LCD at every second. The following command is used to update the LCD:

LCDOUT dec2 hour,":",dec2 minute,":",dec2 second

Bit 2 of register INTCON is cleared at the end of the ISR so that further timer interrupts can be accepted by the microcontroller. Command RESUME terminates the ISR and returns control to the main program.

PROJECT 17 – Interrupt Driven Event Counter

Function

This project demonstrates how we can use the external interrupt input to generate interrupts and then display the number of interrupts occurred on an LCD display. The interrupt input (INT) of a PIC16F84 type microcontroller is normally held HIGH using a pull-up resistor. When this pin goes LOW, an interrupt is generated and the microcontroller increments and displays the total count on an LCD. This project can be used to count the number of external events occurring on the INT pin. For example, a light sensitive sensor can be used to count the number of objects passing on a conveyor belt. Normally the output of the sensor should be LOW. When an object passes in front of the sensor, the sensor should generate a pulse and this pulse can then be used to trigger the INT pin of the microcontroller. The project is based upon the PIC BASIC PRO language.

Circuit Diagram

The circuit diagram of this project is same as in Fig. 5.32, with the addition of a 10K pull-up resistor from pin 6 (RB0/INT) to the +5 V supply. External events should be connected to this pin of the microcontroller. The LCD is connected in the standard way.

Program Description

The program is very simple as most of the LCD functions are included in the PIC BASIC PRO language. The following PDL describes the operation of the program:

Main Program

START

Wait a second until the LCD initializes

Clear the LCD display

Enable external INT interrupts

Define interrupt service routine

Clear count

DO FOREVER

Wait for an interrupt

ENDDO

END

```
'*******************************************************************
'
'        PROJECT:      PROJECT16
'        FILE:         PROJ16.BAS
'        DATE:         August 2000
'        PROCESSOR:    PIC16F84
'        COMPILER:     PIC BASIC PRO
'
'
' This project shows how to use a Timer and also how to create
' and service interrupts from the PIC BASIC PRO language.
'
' A LCD display is connected to port B of the microcontroller.
' In addition, a push button switch, named START is connected
' to pin 2 of port B and another push button switch, named STOP
' to pin 3 of port B.  When START is pressed, the program
' generates timer interrupts every second and then the elapsed
' time is displayed on the LCD in the format "HH:MM:SS".  When
' the STOP button is pressed, the display stops updating.
' Pressing the START button again starts the display timing from
' 00:00:00.
'
'*******************************************************************
'
' Assign LCD pins to PORT B
'
DEFINE LCD_DREG PORTB                    'Using PORT B for LCD data
DEFINE LCD_DBIT 4                        'Using upper 4 bits
DEFINE LCD_RSREG PORTB                   'Using PORT B for RS reg
DEFINE LCD_RSBIT 1                       'Using PORTB.1 for RS pin
DEFINE LCD_EREG PORTB                    'Using PORT B for E reg
DEFINE LCD_EBIT 0                        'Using PORTB.0 for E pin
DEFINE LCD_BITS 4                        'LCD is in 4 bit data mode
DEFINE LCD_LINES 2                       'LCD has 2 lines

hour     var    byte              'hours variable
minute   var    byte              'minutes variable
second   var    byte              'seconds variable
update   var    byte              'display update variable
cnt      var    word              'count
flag     var    byte              'flag

         TRISB = %00001100        'Configure RB2,RB3 as inputs
         OPTION_REG = 2           'Use prescaler 1:8 (8 microsecond)
         ON INTERRUPT GOTO ISR    'Interrupt service routine

         PAUSE 1000               'Wait until the LCD initializes
         LCDOUT $FE,1,"READY"     'Display READY message
```

Fig. 5.42 Program listing of Project 16

```
' Now wait until the START button is pressed.
' Then, initialise the timer interrupt TMR0, clear hours,
' minutes, seconds, and the cnt.
'
strt:
            flag=0
            WHILE flag = 0                  'Do until START pressed
            IF PORTB.2 = 0 THEN             'START Pressed ?
                    TMR0=131                'Set TMR0
                    INTCON=$A0              'Start timer interrupt
                    flag=1                  'flag to exit this routine
                    hour=0                  '
                    minute=0               '
                    second=0               '
                    cnt=0                   '
            ENDIF
            WEND

'
' The following loop is executed continuously.  When update
' is 1 (every second), the LCD display is updated to show the
' new hours, minutes, and seconds.
'
loop:
            IF update = 1 THEN
                    update=0
                    LCDOUT $FE,1
                    LCDOUT dec2 hour,":",dec2 minute,":",dec2 second
            ENDIF
'
' Check if STOP button is pressed.  If so, stop the interrupts.
'
            IF PORTB.3 = 0 THEN
                    INTCON=0
                    GOTO strt               'Go back and wait for START
            ENDIF
            GOTO loop

disable
'
' The following is the interrupt service routine.  Here, the
' second, minute, and the hour variables are updated when a
' second has elapsed.
'
ISR:
            TMR0=131
            cnt=cnt+1
            IF cnt < 1000 THEN noupdate
            update=1
            cnt=0
            second=second+1                          'Update seconds
            IF second = 60 THEN
                    second=0
                    minute=minute+1                  'Update minutes
                    IF minute = 60 THEN
                            minute=0
                            hour=hour+1              'Update hours
                            IF hour=23 THEN
                                    hour=0
                            ENDIF
                    ENDIF
            ENDIF

noupdate:
            INTCON.2=0                       'Clear timer overflow flag
            RESUME                           'Exit interrupt service routine
            ENABLE
END
```

Fig. 5.42   (*Continued*)

Interrupt Service Routine

START

Increment count

Display count on LCD

Return from interrupt

END

```
'****************************************************************
'
'       PROJECT:        PROJECT17
'       FILE:           PROJ17.BAS
'       DATE:           August 2000
'       PROCESSOR:      PIC16F84
'       COMPILER:       PIC BASIC PRO
'
'
' This is an interrupt driven event counter project.
'
' This project shows how to use the external interrupt input
' of the PIC microcontroller.  An LCD display is connected
' to the microcontroller.  The external interrupt pin (INT)
' is normally held at logic HIGH using a pull-up resister.
' An external interrupt occurs on the HIGH to LOW transition
' of a pulse on thin pin.
'
' A light sensor can be connected to the INT pin and the
' project can be used to count objects passing on a conveyor
' belt.  Normally the output of the sensor will be HIGH. When
' an object (e.g. a bottle) passes in front of the object,
' a pulse can be generated and this can interrupt the processor.
' The total number of events occurred will be displayed on a
' LCD.
'
'****************************************************************
i       var byte
cnt     var word
        PAUSE 1000              'wait until the LCD initializes
        OPTION_REG = 0          'interrupt on falling edge of INT
        INTCON = $90            'enable external INT interrupt
        ON INTERRUPT GOTO ISR   'interrupt service routine
        cnt = 0                 'clear count to 0
        LCDOUT $FE,1,"READY"    'display READY message
'
' The following routine loops forever and when an interrupt
' occurs, program jumps to the interrupt service routine,
' named with label ISR.
'
loop:   i = i + 1
        GOTO loop


' This is the interrupt service routine.  The value of the
' count is incremented by 1 and then the total count is
' displayed on the LCD.
'
DISABLE
ISR:
        cnt = cnt + 1           'increment count
        LCDOUT $FE,1,#cnt       'display count
        INTCON.1=0              're-enable INT interrupts
        RESUME                  'return from interrupt
ENABLE

END
```

Fig. 5.43 Program listing of Project 17

Program Listing

The complete program listing is shown in Fig. 5.43. At the beginning of the program there is a 1 second delay so that the LCD initializes properly. Register OPTION_ REG is configured so that interrupts can be recognized on the falling edge of a pulse on pin INT. Register INTCON is configured so that global interrupts and external interrupt INT are enabled. The interrupt service routine is defined to start at label named ISR. The program displays message READY on the LCD and then waits in a loop for interrupts. Note that a dummy instruction is used while waiting so that interrupts can be recognized between statements.

When an interrupt occurs the program jumps to label ISR. Here, variable *cnt* is incremented by 1 and the total count is displayed on the LCD. Bit 1 of register INTCON is cleared (INTCON.1 = 0) so that further external interrupts can be recognized by the microcontroller. The program returns from the interrupt service routine by executing the RESUME instruction.

Chapter 6

Sound Projects

In this chapter we shall be looking at how we can interface our microcontroller to sound generating devices. Sound projects are based on audible devices and these devices have many applications in electronics, ranging from warning devices, burglar alarms, speech processing applications, electronic organs and so on.

Electronic sound generation requires an electronic audible device. There are several such devices available and the details of these devices are given below:

- *Piezo sounders*: These devices operate by an external d.c. source. An internal oscillator applies an a.c. signal to a piezo substrate and this causes alternating deformation of the disk, producing sound output. These devices require about 8 to 20 mA current and generate a sound output of 80 dBA to 100 dBA, at a distance of approximately 30 cm. The frequency response of these devices are in a narrow band, generally in the region 3 kHz to 5 kHz. Piezo sounders usually emit a single tone but some models can emit two or more tones and can also provide pulsed tone outputs. Piezo sounders operate over a wide d.c. voltage range and as a result of this, they are widely used in small portable electronic equipment.

- *Buzzers*: These are mechanical devices which produce sound via a magnetized arm repeatedly striking a diaphragm. These devices operate with a d.c. voltage and the current requirement is small, generally in the region of 10 mA. Buzzers generate a "buzzing" noise (single tone) in the frequency range 300 Hz to 500 Hz. Buzzers are small devices and they can either be panel mounted or PCB mounted.

- *Sounders*: These audible devices generally operate with a d.c. voltage in the range 3 V to 24 V. The current requirement is around 15 mA. The sound output of sounders is single tone at 3 kHz or less, with 80–85 dBA at a distance of 30 cm.

- *Transducers*: These devices generally operate with a small d.c. voltage (around 3 V) and require external drive circuitry. The sound output is 85 dBA or more at a distance of 30 cm. The resonant frequency of transducers is 3 kHz or less. These devices are usually used as mini speakers in PCB mounted applications.

- *Coil type*: These devices operate by a coil attracting and repelling a magnetized diaphragm. The principle of operation is the same as a loudspeaker and in fact these are tiny speakers. An external drive circuit is usually required to generate

119

sound. Coil type audible devices are generally used when it is required to generate multitone sound or speech.

As shown in Fig. 6.1(a), small buzzers which draw several milliamperes of current and operate with around +5 V can directly be connected to the microcontroller I/O ports. Buzzers which require higher currents can be connected to the microcontroller ports using transistor or CMOS switching devices (Fig. 6.1(b)). Small speakers can be connected to the microcontroller I/O ports directly using decoupling capacitors as shown in Fig. 6.1(c). Larger speakers require audio amplifiers before they can be connected to the microcontroller I/O ports (Fig. 6.1(d)).

In this chapter we shall be interfacing our microcontroller to small speakers to see how we can generate different types of sounds.

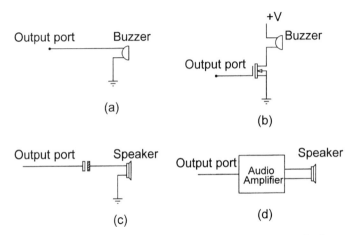

Fig. 6.1   Connecting various audio devices to the microcontroller

PROJECT 18 – Simple Microphone Interface

Function

This project shows how we can interface our microcontroller to a small speaker. The project will send a square wave signal with a frequency of 100 Hz, using a time delay to generate a signal with a period of 10 ms.

Circuit Diagram

The circuit diagram of this project is shown in Fig. 6.2. A PIC16F84 type microcontroller is used for the project and bit 0 of port B (RB0) is connected to a small speaker via a 10 µF electrolytic capacitor.

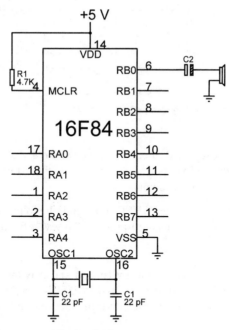

Fig. 6.2 Circuit diagram of Project 18

Program Description

The output of the port is toggled from LOW to HIGH with a time delay of 5 ms between each output. Figure 6.3 shows the output waveform created at the output port. The following PDL describes the operation of the program:

START

> Set port LOW to start with

> **DO FOREVER**

>> Toggle the port status

>> Delay 5 ms

> **ENDDO**

END

Program Listing

The program listing is given in Fig. 6.4. Variable *buzzer* is set to 0 and this variable is used to define the output port pin. Variable *time_delay* is set to 5. Pin 0 of port B is initially set to 0. The port status is then toggled using the command:

TOGGLE buzzer

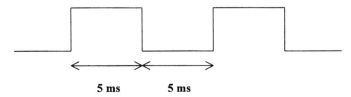

Fig. 6.3 Output waveform of Project 18

The program is repeated after a 5 ms delay. Note that the TRIS command is not required to define the port direction since the LOW command automatically configures the pin as an output.

In general, when using the PAUSE command, the following equation can be used to determine the required delay to produce the given frequency:

$$f = 500/p$$

where f is the required frequency (in Hz) and p is the value used in the PAUSE command. Similarly, we can use the PAUSEUS command (only in PIC BASIC PRO) for microseconds of delay and the relationship between the delay value and the frequency is:

$$f = 500/m$$

where f is the required frequency (in kHz) and m is the value used in the PAUSEUS command. For example, to generate a 1 kHz signal, m = 500 and we have to use the command:

PAUSEUS 500

```
'****************************************************************
'
'         PROJECT:        PROJECT18
'         FILE:           PROJ18.BAS
'         DATE:           August 2000
'         PROCESSOR:      PIC16F84
'         COMPILER:       PIC BASIC
'
'
' A small speaker is connected to pin 0 of port B (RB0). This
' project sends a signal at 100 Hz to the buzzer using a time_delay
' of 5 ms between each pulse.
'****************************************************************

symbol buzzer = 0                       'Assign buzzer to 0
symbol time_delay = 5                   'Assign time_delay to 5

        HIGH buzzer                     'Set buzzer to start with
loop:
        TOGGLE buzzer                   'Toggle buzzer
        PAUSE time_delay                'Delay 5ms
        GOTO loop                       'Go and repeat


        END
```

Fig. 6.4 Program listing of Project 18

Components Required

In addition to the standard microcontroller components, the following components will be required for this project:

C2   10 μF electrolytic capacitor

Small speaker

PROJECT 19 – Using the Built-in SOUND Command

Function

This project shows how we can generate sound using the built-in SOUND command of the PIC BASIC/PRO and the LET BASIC languages. All the possible notes are played on a small speaker connected to the microcontroller.

Circuit Diagram

The circuit diagram of this project is the same as in Fig. 6.2, i.e. a small speaker is connected to port RB0 of the PIC16F84 microcontroller using a decoupling capacitor.

Program Description

The program sounds all the possible notes that can be generated using the SOUND command (PIC BASIC and PIC BASIC PRO only). The SOUND command has the following format:

SOUND Pin,(Note,Duration,Note,Duration, . . .)

The SOUND command generates a square wave type sound. *Note* 0 is silence and notes vary from 1 to 255. *Notes* 1–127 are tones. *Notes* 128–255 are white noise. Tones and white noises are in ascending order, i.e. 1 and 128 are the lowest frequencies, 127 and 255 are the highest frequencies. *Note* 1 is about 78.74 Hz and *note* 127 is about 10 kHz. *Duration* is 0–255 and it determines how long the note should be played in 12 ms increments. For example, if the *duration* is set to 100, the note is played for 1200 ms, or 1.2 seconds. The duration is taken to be 10 ms in this example.

The following PDL describes the operation of the program:

START

Play all notes from 1 to 127 with 10 ms duration

Delay 2 seconds

Play all notes from 128 to 255 with 10 ms duration

END

Program Listing

The program listing is given in Fig. 6.5. Variable *notes* is assigned to location B0. Similarly, variable *twosec* is assigned to 2000 so that we can generate a 2 second delay. A *FOR* loop is used to play all the notes from 1 to 127 with a duration of 10 ms. The program delays for 2 seconds and then plays all the notes from 128 to 255 using another *FOR* loop.

```
'*****************************************************************
'
'           PROJECT:        PROJECT19
'           FILE:           PROJ19.BAS
'           DATE:           August 2000
'           PROCESSOR:      PIC16F84
'           COMPILER:       PIC BASIC
'
'
' A small speaker is connected to pin 0 of port B (RB0). This
' project sounds the buzzer for all notes from 1 to 127 and
' then after 2 seconds delay, from 128 to 255.  Notes duration
' is set to 10 i.e. 120ms with a 4MHz clock.
'*****************************************************************
symbol notes = B0                        'Assign notes to B0
symbol twosec = 2000                     'Assign twosec to 2000

        FOR notes=1 TO 127               'Play all notes 1-127
                SOUND 0,(notes,10)       '
        NEXT notes                       '

        PAUSE twosec                     '2 second delay

        FOR notes = 128 TO 255           'Play all notes 128-255
                SOUND 0,(notes,10)       '
        NEXT notes                       '

        END
```

Fig. 6.5 Program listing of Project 19

Using the LET BASIC Language

Figure 6.6 shows how we can use the LET BASIC language to generate sound on the speaker. *Notes* is defined as a variable using the DIM statement. PORT B is then configured as an output port. A *FOR* loop is used to generate all the sounds from 1 to 127 on pin 0 of port B (B.0). After a delay, notes 128 to 255 are generated using another *FOR* loop.

Using the 8-pin PIC12C672 Microcontroller

PIC12C672 is an 8-pin microcontroller with six I/O ports (GP0 to GP5) which can be programmed for four analog inputs or as five digital I/O and one digital input. The device can be configured to use a 4 MHz internal clock and thus requires no external crystal timing device or capacitors. PIC12C672 contains 2K of program memory and 128 bytes of data memory.

```
REM*************************************************************
REM
REM       PROJECT:        PROJECT19
REM       FILE:           PROJ19-1.BAS
REM       DATE:           August 2000
REM       PROCESSOR:      PIC16F84
REM       COMPILER:       LET BASIC
REM
REM
REM A small speaker is connected to pin 0 of port B.  This project
REM sounds the buzzer for all notes from 1 to 127 and then after
REM 200 milliseconds delay, from 128 to 255.
REM *************************************************************

DEVICE 16F84
DIM notes
DEFINE PORTB=00000000

        FOR notes=1 TO 127
                SOUND(notes,100,B.0)
        NEXT notes

        DELAYMS(200)

        FOR notes=128 TO 255
                SOUND (notes,100,B.0)
        NEXT notes

        END
```

Fig. 6.6 LET BASIC program listing of Project 19

Figure 6.7 shows how we can connect a small speaker to port GP0 of the microcontroller. Notice that the microcontroller has been configured to use its internal RC timing (during programming of the device) and there is no crystal or timing capacitors. Also, the external reset resistor on MCLR pin (pin 4) is generally not required and is not used here. The program listing in Fig. 6.8 shows how we can program the PIC12C672 microcontroller using the PIC BASIC PRO language. *Notes* is defined as a byte variable and this variable is used in the *FOR* loops. Variable *prt* is assigned to port GP0 (programmed as GPIO.0) of the microcontroller where the speaker is connected. Register ADCON1 is set to 7 so that all the microcontroller ports operate as digital I/O. A *FOR* loop generates all the notes from 1 to 127 with a duration of 10 ms. Note that square brackets are used in the SOUND

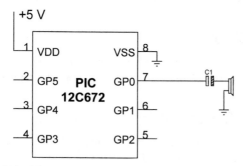

Fig. 6.7 Using a PIC12C672 microcontroller

```
'****************************************************************
'
'          PROJECT:          PROJECT19
'          FILE:             PROJ19-2.BAS
'          DATE:             August 2000
'          PROCESSOR:        PIC12C672
'          COMPILER:         PIC BASIC PRO
'
'
' A small speaker is connected to pin 0 of port B (RB0).  The
' project sounds the buzzer for all notes from 1 to 127 and
' then after 2 seconds delay, from 128 to 255.  Notes duration
' is set to 10 i.e. 120ms with a 4MHz clock.
'****************************************************************

notes var byte                          'Assign notes to B0
prt var GPIO.0
twosec con 2000                         'Assign twosec to 2000

        ADCON1=7                        'Configure ports as digital

        FOR notes=1 TO 127               'Play all notes 1-127
                SOUND prt,[notes,10]     '
        NEXT notes                       '

        PAUSE twosec                    '2 second delay

        FOR notes = 128 TO 255          'Play all notes 128-255
                SOUND prt,[notes,10]     '
        NEXT notes                       '

        END
```

Fig. 6.8 Program listing for the 12C672

command when we program with the PIC BASIC PRO language. After a 2 second delay, all the notes from 128 to 255 are played with a 10 ms duration.

PROJECT 20 – Using the Built-in FREQOUT Command

Function

This project shows how we can generate sound using the FREQOUT command of the PIC BASIC PRO language. In this project, all the eight notes of a musical octave are played with a 1 second delay between each note.

Circuit Diagram

The circuit diagram of this project is the same as in Fig. 6.2, i.e. a small speaker is connected to port RB0 of the PIC16F84 microcontroller using a decoupling capacitor.

Program Description

The program generates all the eight notes of a musical octave, delays for 2 seconds and then repeats itself. The following octave is used for this project (the frequencies are in Hz):

Notes:	C	D	E	F	G	A	B	C
Freq:	262	294	330	349	392	440	494	524

The following PDL describes the operation of the program:

START

    DO FOREVER

        Generate a 262 Hz signal with 1 second duration

        Generate a 294 Hz signal with 1 second duration

        Generate a 330 Hz signal with 1 second duration

        Generate a 349 Hz signal with 1 second duration

        Generate a 392 Hz signal with 1 second duration

        Generate a 440 Hz signal with 1 second duration

        Generate a 494 Hz signal with 1 second duration

        Generate a 524 Hz signal with 1 second duration

        Delay 2 seconds

    ENDDO

END

Program Listing

The full program listing is shown in Fig. 6.9. Variable *buzzer* is assigned to pin 0 of port B. Variable *twosec* is assigned to number 2000 so that we can generate a 2 second delay. Command FREQOUT is used to generate a sound with a specified frequency. Note that this command is only available under the PIC BASIC PRO language and its general format is as follows:

    FREQOUT Pin,Onms,Frequency1,Frequency2 . . .

which produces sounds with frequency1, frequency2 and so on, with a duration of Onms milliseconds. FREQOUT generates tones using a pulse width modulation and the data coming out of the pin is not clean. Some kind of filter is recommended to smooth the signal to a sine wave and to get rid of the harmonics. FREQOUT works best with high oscillator frequency (e.g. 20 MHz). Using smaller frequencies requires better filtering.

More Efficient Program

Figure 6.10 shows a more efficient program which produces the same results as the program in Fig. 6.9. Here, variable notes are declared as a byte array and this array

```
'******************************************************************
'
'          PROJECT:       PROJECT20
'          FILE:          PROJ20.BAS
'          DATE:          August 2000
'          PROCESSOR:     PIC16F84
'          COMPILER:      PIC BASIC PRO
'
'
' A small speaker is connected to pin 0 of port B (RB0).
' This project sounds the buzzer for the following octave
' of notes (the frequencies in Hz are given under the notes):
'
'        C   D   E   F   G   A   B   C
'       262 294 330 349 392 440 494 524
'
' The duration of each note is 1 set to 1 second.  The notes
' are repeated after a 2 second delay.
'******************************************************************
buzzer var PORTB.0                      'Assign buzzer to PORTB.0
symbol twosec = 2000

loop:
        FREQOUT buzzer,1000,262         'Freq=262 Hz
        FREQOUT buzzer,1000,294         'Freq=294 Hz
        FREQOUT buzzer,1000,330         'Freq=330 Hz
        FREQOUT buzzer,1000,349         'Freq=349 Hz
        FREQOUT buzzer,1000,392         'Freq=392 Hz
        FREQOUT buzzer,1000,440         'Freq=440 Hz
        FREQOUT buzzer,1000,494         'Freq=494 Hz
        FREQOUT buzzer,1000,524         'Freq=524 Hz

        PAUSE twosec                    'Delay 2 seconds
        GOTO loop                       'Go back and repeat

        STOP
        END
```

Fig. 6.9 Program listing of Project 20

stores all the frequencies that we wish to output. A FOR loop is used to send each frequency to the output using the FREQOUT command. Different frequencies are selected by varying the index of array notes, i.e. the following command:

FREQOUT buzzer,1000,notes[J]

generates a sound with a frequency of notes[J], and a duration of 1 second on the pin buzzer.

PROJECT 21 – Simple Electronic Organ

Function

This is a simple electronic organ project. A small speaker is connected to bit 0 of port A. Eight push-button switches are connected to port B to act as the keyboard for the electronic organ. Only one octave (eight notes) is provided.

```
'*************************************************************
'
'        PROJECT:        PROJECT20
'        FILE:           PROJ20-1.BAS
'        DATE:           August 2000
'        PROCESSOR:      PIC16F84
'        COMPILER:       PIC BASIC PRO
'
'
' A small speaker is connected to pin 0 of port B (RB0).
' This project sounds the buzzer for the following octave
' of notes (the frequencies in Hz are given under the notes):
'
'        C    D    E    F    G    A    B    C
'        262  294  330  349  392  440  494  524
'
' The duration of each note is 1 set to 1 second.  The notes
' are repeated after a 2 second delay.
'*************************************************************
buzzer var PORTB.0                       'Assign buzzer to PORTB.0
symbol twosec = 2000
notes var word[8]
j var byte

notes[0] = 262  : notes[1] = 294  : notes[2] = 330  : notes[3] = 349
notes[4] = 392  : notes[5] = 440  : notes[6] = 494  : notes[7] = 524

loop:
        FOR J = 0 TO 7
                FREQOUT buzzer,1000,notes[J]
        NEXT J

        PAUSE twosec                     'Delay 2 seconds
        GOTO loop                        'Go back and repeat

        END
```

Fig. 6.10 More efficient program for Project 20

Circuit Diagram

The circuit diagram of this project is shown in Fig. 6.11. The speaker is connected to pin 0 of port A. The keyboard switches are connected to port B. Bit 0 is assigned to note C, bit 1 is assigned to note D, bit 2 is assigned to note E and so on. The switches are normally held at logic HIGH with pull-up resistors. Pressing a switch sends a logic LOW to the microcontroller port.

Program Description

In this project, the following octave of notes is used:

Switch:	1	2	3	4	5	6	7	8
Notes:	C	D	E	F	G	A	B	C
Freq:	262	294	330	349	392	440	494	524

The program continuously checks the switches and if any switch is pressed then the note corresponding to that switch position is sent to the speaker. The following PDL describes the operation of the program:

START

> Configure port B as inputs

> Configure port A as outputs

> **DO FOREVER**

>> **IF** a key is pressed **THEN**

>>> Get key number

>>> Send note corresponding to this key to the speaker

>> **ENDIF**

> **ENDDO**

END

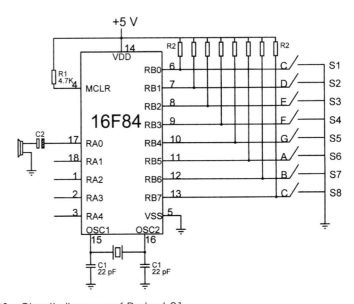

Fig. 6.11 Circuit diagram of Project 21

Program Listing

The full program listing is shown in Fig. 6.12. Variable *buzzer* is assigned to pin 0 of port A (PORTA.0). Variable *key_pressed* is defined as a byte. Variable *notes* is declared as an array with eight elements, from *notes[0]* to *notes[8]*. The frequencies of the musical octave are assigned to the array elements. Port B is then configured as an input port and port A is configured as an output port. Switch status is checked by reading data from port B and then comparing it with decimal number 255:

    IF PORTB <> 255 THEN

```
'****************************************************************
'
'          PROJECT:        PROJECT21
'          FILE:           PROJ21.BAS
'          DATE:           August 2000
'          PROCESSOR:      PIC16F84
'          COMPILER:       PIC BASIC PRO
'
'
' This is a simple electronic organ project.  8 switches are
' connected to port B of the microcontroller.  In addition,
' a small speaker is connected to pin 0 of port A.  Musical
' notes are played when the keys are pressed.

' This project sounds the speaker for the following octave
' of notes (the frequencies are given under the notes):
'
'        C   D   E   F   G   A   B   C
'       262 294 330 349 392 440 494 524
'
'****************************************************************
      buzzer var PORTA.0                    'Assign buzzer to PORTA.0
      notes var word[9]
      key var byte
      key_pressed var byte

      notes[1] = 262  : notes[2] = 294  : notes[3] = 330  : notes[4] = 349
      notes[5] = 392  : notes[6] = 440  : notes[7] = 494  : notes[8] = 524

              TRISB = %11111111
              TRISA = 0
      loop:
              IF PORTB <> 255 THEN          'Check if any key pressed
                      key = ~PORTB          'Invert key data
                      key_pressed = NCD key
                      FREQOUT buzzer,5,notes[key_pressed]
              ENDIF

              GOTO loop                     'Go back and repeat

              END
```

Fig. 6.12 Program listing of Project 21

Decimal number 255 has the bit pattern "11111111" and this is the normal state of port B when no keys are pressed since the port pins are pulled high with resistors. Thus the commands within the *IF* block are not normally executed. When a key is pressed, the above condition becomes true (i.e. PORT B is not equal to all 1s) and the commands within the *IF* block are executed. The status of port B is inverted and the number obtained is the bit position pressed by the user. Bit positions corresponding to switches 1, 2, 3, 4, 5, 6, 7 and 8 are 1, 2, 4, 8, 16, 32, 64 and 128 respectively. For example, if the user presses key 5, number 16 will be obtained and assigned to variable *key_pressed* as shown below:

Normal state of port B: 1 1 1 1 1 1 1 1

State when key 5 is pressed: 1 1 1 0 1 1 1 1

State when the port data is inverted: 0 0 0 1 0 0 0 0

Command NCD (encode) of the PIC BASIC PRO language is then used to convert this number into a number which shows the bit position within the number. For example, if variable x is 16, the command:

    y = NCD x

will return 5 in variable y, i.e. bit position 5 is set in variable x. This value is then used as an index in array notes and the FREQOUT command is used to send the correct note to the speaker:

    FREQOUT buzzer,5,notes[key_pressed]

The note is sounded for a duration of 5 ms.

Components Required

In addition to the standard microcontroller components, the following components will be required for this project:

    C2   10 µF electrolytic capacitor

    R2   10K resistors (8 off)

    S   Push-button switches (8 off)

          Small speaker

Chapter 7

Temperature Projects

Temperature measurement and control is one of the most common applications of microcontroller-based data acquisition systems. Four types of sensors are commonly used to measure temperature in commercial and industrial applications. These are *thermocouples, resistive temperature devices* (RTDs), *thermistors*, and *integrated circuit* (IC) *temperature sensors*. Each sensor has its unique advantages and disadvantages and by understanding how these sensors work, and what types of signal conditioning are required for each, we can make more accurate and reliable temperature measurement, monitoring, and control.

The typical characteristics of various temperature sensors are described briefly below.

Thermocouples: These are inexpensive sensors which have a wide range of temperature range. Thermocouples work on the principle that when two dissimilar metals are combined, a voltage appears across the junction between the metals. By measuring this voltage, we can get a temperature reading. Different combinations of metals create different thermocouple voltages and there is a wide range of thermocouples available for different applications. Thermocouples generate very low voltages, typically $50\,\mu V/°C$. These low-level signals require special signal conditioning to remove any possible noise. Thermocouples have non-linear relationships to the measured temperature and as a result of this it is necessary either to linearize the characteristics or to use look-up tables to obtain the actual temperature from the measured voltage. Analogue-to-digital converter devices are required to connect the thermocouples to computer-based equipment.

RTDs: An RTD is a resistor with its resistance changing with temperature. The most popular type of RTD is made of platinum and has a resistance of $100\,\Omega$ at 0°C. Because RTDs are resistive devices, a current must pass through the RTD to produce a voltage that can be measured. The change in resistance is very small (about $0.4\,\Omega/°C$) and special circuitry is generally needed to measure the small changes in temperature. One of the drawbacks of RTDs is their non-linear change in resistance with temperature. RTDs are analogue devices and analogue-to-digital converters are required to interface these devices to computers.

Thermistors: Thermistors are metal oxide semiconductor devices whose resistance changes with temperature. One of the advantages of thermistors is their fast responses and high sensitivity. For example, a typical thermistor may have a resistance of $50\,k\Omega$ at 25°C, but have a resistance of only $2\,k\Omega$ at 85°C. Like RTDs, a current is passed through a thermistor and the voltage across the thermistor is measured. Thermistors are very non-linear devices and look-up tables are usually used to convert the measured voltage to temperature. Thermistors are very small devices and one downside of this is that they can be self-heating under a large excitation current. This of course increases the temperature of the device and can give erroneous results. Thermistors are analogue sensors and analogue-to-digital converters are required to interface these sensors to computer-based equipment.

IC Temperature Sensors: Integrated circuit temperature sensors are usually 3- or 8-pin active devices which require a power supply to operate and they give out a voltage which is directly proportional to the temperature. There are basically two types of IC temperature sensors: analogue sensors are usually 3-pin devices and they give out an analogue voltage of typically $10\,mV/°C$ which is directly proportional to the temperature; digital temperature sensors provide 8- or 9-bit serial digital output data which is directly proportional to the temperature.

In this chapter we shall be looking at how we can interface various temperature sensors to PIC microcontrollers in order to measure and display the ambient temperature.

PROJECT 22 – Using a Digital Temperature Sensor

Function

This project shows how we can interface a DS1620 type digital temperature sensor to a PIC16F84 type microcontroller. In this project the ambient temperature is measured continuously and then displayed in degrees centigrade on three TIL311 type alphanumeric displays. Positive temperature is displayed from 0°C to 125°C. Negative temperature is displayed with a leading letter "E" in the range down to −55°C.

Circuit Diagram

The block diagram of this project is shown in Fig. 7.1. DS1620 is a digital IC temperature sensor which measures the ambient temperature and provides the output as 9 bits of digital serial data. The microcontroller extracts the temperature data from the DS1620 and then displays the temperature on three TIL311 type alphanumeric displays.

Before describing the circuit diagram in detail, it is useful to look at the operation of the DS1620 temperature sensor IC.

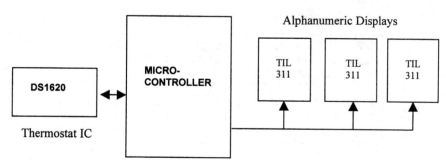

Fig. 7.1 Block diagram of Project 22

DS1620 Digital Thermometer IC

DS1620 is a digital thermometer and thermostat IC which provides 9 bits of serial data to indicate the temperature of the device. The pin configuration of the DS1620 is shown in Fig. 7.2. VDD is the power supply which is normally connected to a +5 V supply. DQ is the data input/output pin. CLK is the clock input. RST is the reset input. The device can also act as a thermostat. THIGH is driven high if the DS1620's temperature is greater than or equal to a user defined

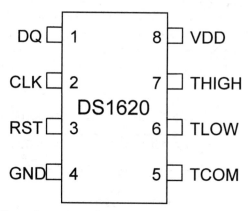

Fig. 7.2 Pin configuration of DS1620

temperature TH. Similarly, TLOW is driven high if the DS1620's temperature is less than or equal to a user defined temperature TL. TCOM is driven high when the temperature exceeds TH and stays high until the temperature falls below TL. User defined temperatures TL and TH are stored in non-volatile memory of the device so that they are not lost even after removal of the power.

Data is output from the device as 9 bits, with the LSB sent out first. The temperature is provided in two's complement format from −55°C to +125°C, in

Table 7.1 Temperature/data relationship of DS1620

TEMP (°C)	Digital Output (Binary)	Digital Output (Hex)	2's complement	Digital Output (Decimal)
+125	0 11111010	0FA	–	250
+25	0 00110010	032	–	50
0.5	0 00000001	001	–	1
0	0 00000000	000	–	0
−0.5	1 11111111	1FF	001	511
−25	1 11001110	1CE	032	462
−55	1 10010010	192	06E	402

steps of 0.5°C. Table 7.1 shows the relationship between the temperature and data output by the device.

Operation of DS1620

Data input and output is through the DQ pin. When RST input is high, serial data can be written or read by pulsing the clock input. Data is written or read from the device in two parts. First, a protocol is sent and then the required data is read or written. The protocol is an 8-bit data and the protocol definitions are given in Table 7.2. For example, to write the thermostat value TH, the hexadecimal

Table 7.2 DS1620 protocol definitions

PROTOCOL	PROTOCOL DATA (Hex)
Write TH	01
Write TL	02
Write Configuration	0C
Stop Conversion	22
Read TH	A1
Read TL	A2
Read Temperature	AA
Read Configuration	AC
Start Conversion	EE

protocol data 01 is first sent to the device. After issuing this command, the next nine clock cycles clock in the 9-bit temperature limit which will set the threshold for operation of the THIGH output.

For example, the following data (in hexadecimal) should be sent to the device to set it for a TH limit of +50°C and a TL limit of +20°C and then subsequently to start the conversion:

01 Send TH protocol

64 Send TH limit of 50 (64 hex = 100 decimal)

02 Send TL protocol

28 Send TL limit of 20 (28 hex = 40 decimal)

EE Send conversion start protocol

A configuration/status register is used to program various operating modes of the device. This register is written with protocol 0C (hex) and the status is read with protocol AC (hex). Some of the important configuration/status register bits are as follows:

Bit 0 This is the 1 shot mode. If this bit is set, the DS1620 will perform one temperature conversion when the start convert protocol is sent. If this bit is 0, the device will perform continuous temperature conversions.

Bit 1 This bit should be set to 1 for operation with a microcontroller or microprocessor.

Bit 5 This is the TLF flag and is set to 1 when the temperature is less than or equal to the value TL.

Bit 6 This is the THF bit and is set to 1 when the temperature is greater than or equal to the value of TH.

Bit 7 This is the DONE bit and is set to 1 when a conversion is complete.

The complete circuit diagram of this project is shown in Fig. 7.3. Bit 5 of port B is connected to the RST input of DS1620, bit 6 is connected to the clock input and bit 7 of port B is connected to the DQ pin of the DS1620. Three TL311 type alphanumeric displays are connected to port B of the microcontroller. Digit 1 is controlled from bit 0 of port A, digit 2 from bit 1 of port A, and digit 3 from bit 2 of port A.

Program Description

The program reads the temperature from the DS1620 thermometer IC and displays the temperature on three TIL311 type displays continuously with a 1 second delay between each displayed output. The following PDL describes the operation of the program:

Main Program

START

    Configure DS1620

DO FOREVER

    Read temperature

    Display temperature

    Delay a second

ENDDO

END

Subroutine Configure DS1620

START

    Send protocol "Write Configuration"

    Set configuration/status register to 2 (i.e. continuous operation)

END

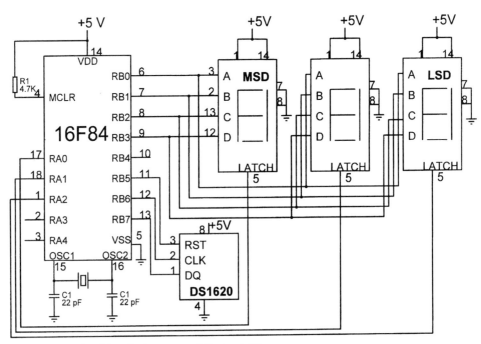

Fig. 7.3 Circuit diagram of Project 22

Subroutine Read Temperature

START

> Send protocol "Read Temperature"
>
> Call subroutine read_from_ds1620 to get the temperature

END

Subroutine Display Temperature

START

> **IF** temperature is negative **THEN**
>
> > Get 2's complement of the temperature reading
> >
> > Divide temperature by 2 to get real temperature
> >
> > Set digit 1 to display letter "E"
> >
> > Display temperature digits
>
> **ELSE**
>
> > Divide temperature by 2 to get real temperature
> >
> > Display temperature digits
>
> **ENDIF**

END

Program Listing

The full program listing is given in Fig. 7.4. DS1620 pins DQ, CLK and RST are assigned to port B pins 7, 6 and 5 respectively. DS1620 protocols are then defined using symbols. Various other symbols and port addresses are then defined at the beginning of the program. The main program is very short and consists of a small loop where the temperature is read from the DS1620 and then displayed using subroutines. The loop is repeated with a 1 second delay between each output.

Subroutine CONFIGURE_DS1620 initializes and configures the DS1620. *Write Configuration* protocol is first sent to the device and then the configuration/status register is loaded with 2 to indicate that we are connecting the DS1620 to a microprocessor/microcontroller. Variable *this_bit* stores the data to be sent to the sensor and data is sent in serial format by calling subroutine WRITE_DS1620_BIT which sends a single bit every time it is called. The data is shifted right in a *FOR* loop and all the 8 bits are sent by calling subroutine WRITE_DS1620_BIT. After sending the protocol message, data 2 is sent using the same programming technique. DS1620 is then put into conversion mode by sending the protocol message *Start Conversion*.

```
'****************************************************************
'
'      PROJECT:        PROJECT22
'      FILE:           PROJ22.BAS
'      DATE:           August 2000
'      PROCESSOR:      PIC16F84
'      COMPILER:       PIC BASIC
'
'
' This is a digital temperature sensor project.  A DS1620 type
' digital thermometer is used to read the ambient temperature.
' The temperature is then displayed on three TIL311 type
' alphanumeric displays.  The temperature range is -55C to +125C.
' Positive temperature is displayed with leading zeroes.
' Negative temperatures is displayed by inserting the letter
' "E" in front of the display.  The display accuracy is +/- 1C
' i.e. there is no decimal point in the displayed data.
'
' The four data inputs (A,B,C,D) of TIL311 displays are connected
' to port pins RB0-RB3.  The latch inputs of the displays are
' conencted to RA0-RA2, with RA0 connected to the MSD device.
' The DS1620 temperature IC is connected to port pins RB5-RB7.
' DQ input is connected to RB7, CLK input is connected to RB6,
' and RST input is connected to RB5.
'
' The display is updated every second.
'
'****************************************************************
' Define DS1620 pins
symbol  ds1620_dq  = Pin7            'DS1620 DQ pin
symbol  ds1620_clk = Pin6            'DS1620 CLK pin
symbol  ds1620_rst = Pin5            'DS1620 RST pin
'
' Define DS1620 functions
symbol  write_config = $0C           'DS1620 write conf
symbol  start_conv   = $EE           'DS1620 start conversion
symbol  read_temp    = $AA           'DS1620 read temperature
'
' Define other symbols
symbol  I = B1
symbol  ds1620_data = W3
symbol  this_bit = W4
symbol  first = B2                   'Display MSD data
symbol  second = B3                  'Display middle data
symbol  third = B4                   'Display LSD data
symbol  temp = B5
symbol  new_bit = W5
'
' Define PORT addresses and directions
symbol  PORTA = 5
symbol  TRISA = $85
symbol  PORTB = 6
symbol  TRISB = $86
'
' Start of MAIN program
'
        POKE TRISA,0                 'Configure port A
        POKE TRISB,0                 'Configure port B

        GOSUB CONFIGURE_DS1620       'Configure DS1620
loop:
        GOSUB READ_TEMPERATURE       'Read temperature
        GOSUB DISPLAY_TEMPERATURE    'Display temperature
        PAUSE 1000                   'Delay a second
        GOTO loop                    'Go back and repeat


'
' End of MAIN program
'
'
```

Fig. 7.4 Full program listing of Project 22

```
' Subroutine CONFIGURE_DS1620.
' This subroutine configures the DS1620 by first writing a
' protocol message and then the actual configuration data.
' The configuration/status register is loaded with 2 and then
' a start conversion protocol message is sent to the DS1620.
'
CONFIGURE_DS1620:
'
' Send "Write Configuration" protocol to DS1620
'
        ds1620_rst  = 1                 'Set RST pin
        this_bit=write_config           'Protocol=write_config
        FOR I = 1 TO 8                  'Send protocol
             GOSUB WRITE_DS1620_BIT     '
             this_bit=this_bit/2        'Shift right
        NEXT I
'
' Load configuration/status register with 2 to say that
' we are connecting the DS1620 to a microprocessor or a
' microcontroller.
'
        ds1620_data = 2                 'Configuration=2
        this_bit=ds1620_data            '
        FOR I = 1 TO 8                  'Send configuration
             GOSUB WRITE_DS1620_BIT     '
             this_bit=this_bit/2        'Shift right
        NEXT I

        ds1620_rst = 0                  'Clear RST pin
'
' Now send "Start Conversion" protocol to DS1620
'
        ds1620_rst = 1                  'Set ds1620_rst=1
        this_bit=start_conv             'Protocol=start_conv
        FOR I = 1 TO 8                  'Send protocol
             GOSUB WRITE_DS1620_BIT     '
             this_bit=this_bit/2        'Shift right
        NEXT I

        ds1620_rst = 0                  'Clear RST pin
        RETURN                          'End of DS1620 configuration


'
' This subroutine sends a bit to the DS1620.  The bit is in
' location this_bit to start with.  This subroutine is used
' by the other subroutines.
'
WRITE_DS1620_BIT:
        ds1620_dq = this_bit            'Get bit to write
        ds1620_clk = 0                  'Set CLK LOW
        ds1620_clk = 1                  'Set CLK HIGH
        ds1620_dq = 1                   'Set DQ HIGH
        RETURN                          'End of write bit

'
' This subroutine reads a bit from the DS1620.  The bit read is
' stored in location this_bit.  This subroutine is used by the
' other subroutines.
'
READ_DS1620_BIT:
        ds1620_clk = 0                  'Clear CLK
        this_bit = Pin7                 'Get bit from DS1620
        ds1620_clk = 1                  'Set CLK
        RETURN                          'End of read bit


'
' This subroutine reads the current temperature from the DS1620.
' First of all, a "Read Temperature" protocol is sent to the
' DS1620.  9 bits of data is then read into location ds1620_data.
'
```

Fig. 7.4 (*Continued*)

```
READ_TEMPERATURE:
'
' Send Read Temperature protocol
'
        ds1620_data = 0
        ds1620_rst = 1                          'Set RST bit
        this_bit=read_temp                      'Protocol=read_temp
        FOR I = 1 TO 8                          'Send protocol
                GOSUB WRITE_DS1620_BIT          '
                this_bit=this_bit/2             'Shift right
        NEXT I
'
' Now read the temperature into ds1620_data.
'
        ds1620_data=0
        new_bit=1
        FOR I=1 TO 9
                GOSUB READ_DS1620_BIT           'Get bits
                this_bit=new_bit * this_bit
                ds1620_data=ds1620_data + this_bit
                new_bit=2 * new_bit
        NEXT I
        ds1620_rst = 0
        RETURN


' This subroutine displays the temperature on the 3 TIL311 type
' displays.  If the temperature is negative, program jumps to
' label temp_negative.
'
DISPLAY_TEMPERATURE:
        IF ds1620_data > 255 then temp_negative
        ds1620_data = ds1620_data / 2           'Extract actual temp
        first = ds1620_data / 100               'Extract MSD digit
        temp = ds1620_data // 100               '
        second = temp / 10                      'Extract middle digit
        third = temp // 10                      'Extract LSD digit

display:
'
' Now display the temperature in variables first,second,third.
'
        POKE PORTA,7                            'Set latches=1
        POKE PORTB,first                        'Get MSD digit
        POKE PORTA,6                            'Clock the latch
        POKE PORTA,7                            '

        POKE PORTB,second                       'Get the middle digit
        POKE PORTA,5                            'Clock the latch
        POKE PORTA,7                            '

        POKE PORTB,third                        'Get the LSD digit
        POKE PORTA,3                            'Clock the latch
        POKE PORTA,7                            '
        RETURN                                  'End of display


' We jump here if the temperature is negative.
'
TEMP_NEGATIVE:
        ds1620_data = ds1620_data ^ $FF         'Complement temp
        ds1620_data=ds1620_data+1               'Add 1 for 2s comp
        ds1620_data=ds1620_data & $FF           'Extract lower byte
        ds1620_data=ds1620_data / 2             'Get actual temp

        first=14                                'Display leading "E"
        second=ds1620_data / 10                 'Extract middle digit
        third=ds1620_data // 10                 'Extract LSD digit
        GOTO display                            'Go and display
END
```

Fig. 7.4 (*Continued*)

Subroutine READ_DS1620_BIT reads a single bit data from the DS1620. The data is returned in variable *this_bit*, one bit at a time.

Subroutine READ_TEMPERATURE reads a temperature sample from the sensor. Protocol *Read Temperature* is first sent to the device. 9 bits of temperature data are then read and stored in variable *ds1620_data*. Note that subroutine READ_ DS1620_BIT is called to read a bit and each bit is shifted left by multiplying by 2 and adding to *ds1620_data*. On return from subroutine READ_TEMPERATURE, variable *ds1620_data* stores the 9-bit temperature data.

Subroutine DISPLAY_TEMPERATURE displays the temperature on the three TIL311 alphanumeric displays. At the beginning of the subroutine the code checks to see if the temperature is negative (temperature > 255) and jumps to label *temp_ negative* to display negative temperatures. Variable *first* stores the MSD digit of the display, variable *second* stores the middle digit, and variable *third* stores the LSD digit of the temperature. These digits are extracted by dividing the temperature using the divide operator ("/") and the remainder ("//") operator. For example, if variable y contains the decimal value 125 then the three digits can be extracted as follows:

first	= y/100	first	= 1
temp	= y//100	temp	= 25 (remainder)
second	= temp/10	second	= 2
third	= temp//10	third	= 5 (remainder)

Note that the PIC BASIC PRO language offers the command DIG which can be used to extract the digit of a number. For example, for the above example, we can use the DIG command as follows:

first   = y DIG 2

second = y DIG 1

third   = y DIG 0

Positive temperatures are displayed with leading zeros. Negative temperatures are displayed by inserting the character "E" at the MSD digit position (decimal number 14 is sent to this digit).

Components Required

In addition to the standard minimum microcontroller components, the following will be required for this project:

TIL311   Alphanumeric displays (3 off)

DS1620   Digital temperature sensor IC

PROJECT 23 – Using Analogue Temperature Sensor IC with Built-in A/D Converter

Function

This project shows how we can connect an analogue temperature sensor IC to a PIC16C71 type microcontroller and then display the temperature. In this project, LM35DZ type analogue temperature IC is connected to one of the analogue ports of the PIC16C71 and temperature is converted to digital and displayed on three TIL311 type alphanumeric displays every second. The block diagram of this project is shown in Fig. 7.5.

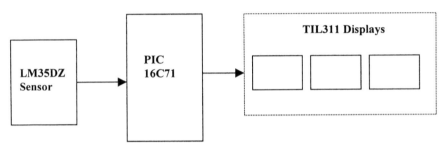

Fig. 7.5   Block diagram of Project 23

Circuit Diagram

The circuit diagram of Project 23 is shown in Fig. 7.6. Three TIL311 type alphanumeric displays are connected to port B of a PIC16C71 microcontroller. Data inputs of the displays (A, B, C, D) are connected to ports RB0–RB3. Display latches are controlled by the upper nibble of port B. Pin RB7 is connected to the first display (MSD) digit latch, pin RB6 is connected to the middle digit display latch, and pin RB5 is connected to the LSD digit display latch.

An LM35DZ type analogue temperature sensor IC is connected to port RA0 of the microcontroller. LM35DZ is a simple but accurate 3-pin temperature sensor IC. Pin 1 of the device is connected to the power supply (+5 V), pin 3 is connected to the ground. Pin 2 is the output and this output provides a voltage which is directly proportional to the ambient temperature. The device can measure temperature from 2°C up to 100°C (some types can measure a wider range) and the output voltage to temperature relationship is 10 mV/°C. For example, at 20°C the output is 200 mV. Similarly, at 35°C, the output voltage is 350 mV, and so on.

Program Description

The program is based upon the PIC BASIC language. At the beginning of the program the I/O ports and the A/D converter of the PIC16C71 is configured.

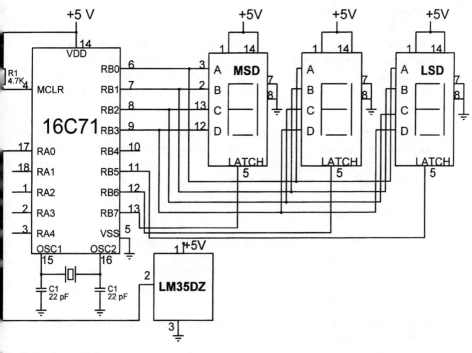

Fig. 7.6 Circuit diagram of Project 23

Temperature is then read from the LM35DZ and converted into digital form. This temperature is displayed on the alphanumeric displays after being scaled appropriately. The process is repeated every second. The following PDL describes the operation of the program:

START

      Configure I/O ports

      Configure A/D converter

    DO FOREVER

      Read temperature from LM35DZ

      Convert to digital form

      Scale the temperature

      Display the temperature

    ENDDO

  END

Program Listing

The complete program listing is shown in Fig. 7.7. Variables *latch_msd*, *latch_middle*, and *latch_lsd* are the three display latches and these are assigned to pins 7, 6 and 5 of port B respectively. Port A and port B data and direction register addresses are then defined. Port A is set as an input port and port B is set as an output port.

Register ADCON1 of the microcontroller is cleared to 0 so that all the port A pins are configured as analogue inputs. The A/D converter is then configured to use a clock frequency of $f_{OSC/8}$ and the A/D converter is turned on by sending the bit pattern "01000001", i.e. the command:

POKE ADCON0,$41

The program then enters a loop where the temperature is converted to digital and then displayed. The command:

POKE ADCON,$45

starts the A/D converter on channel 0 of the microcontroller. The conversion status is checked by testing bit 1 of ADCON1:

wait: PEEK ADCON0,B0

IF Bit1 = 0 THEN wait

```
'****************************************************************
'
'          PROJECT:        PROJECT23
'          FILE:           PROJ23.BAS
'          DATE:           August 2000
'          PROCESSOR:      PIC16C71
'          COMPILER:       PIC BASIC
'
'
' This is a simple analogue temperature sensor project.
' An LM35DZ type analogue temperature sensor is connected
' to port RA0 of a PIC 16C71 type microcontroller.  In
' addition, 3 TIL311 type alphanumeric displays are connected
' to port B of the microcontroller.
' The program reads the analogue temperature, converts it to
' digital form and then displays on the TIL311 displays every
' second.
'
'****************************************************************
'
'Define the display latches
symbol latch_msd = Pin7              'MSD digit latch
symbol latch_middle = Pin6           'Middle digit latch
symbol latch_lsd = Pin5              'LSD digit latch

symbol new_data = B5                 'A/D converter data
'
' Define port addresses
```

Fig. 7.7 Program listing of Project 23

```
        symbol PORTA = 5
        symbol TRISA = $85
        symbol PORTB = 6
        symbol TRISB = $86
        '
        ' Define A/D converter registers
        symbol ADCON0 = 8                       'ADCON0 register
        symbol ADRES = 9                        'ADRES register
        symbol ADCON1 = $88                     'ADCON1 register
        '
        ' Define display registers
        symbol first = B1                       'MSD digit
        symbol second = B2                      'Middle digit
        symbol third = B3                       'LSD digit
        '
        ' Define other variables
        symbol temp = B4
        symbol secs = 1000                      'seconds delay
        '
        ' Configure port A and B
                POKE TRISA,%11111111            'Configure as inputs
                POKE TRISB,0                    'Configure as outputs

                POKE ADCON0,0                   'Select RA0-RA3
                POKE ADCON0,$41                 'Set A/D oscillator

loop:
                POKE ADCON0,$45                 'Start conversion
wait:           PEEK ADCON0,B0                  'Conversion complete ?
                IF Bit1 = 0 THEN wait           '
                PEEK ADRES,new_data             'Get A/D data

                GOSUB DISPLAY_TEMPERATURE       'Display data
                PAUSE secs                      'Delay a second
                GOTO loop                       'Go back and repeat

        '
        ' This subroutine displays the data on 3 TIL311 type
        ' alphanumeric displays.  The converted data (new_data) is first
        ' multiplied by 2 to get the true temperature in C̄.
        ' Variables first, second, and third store the data to be
        ' displayed on the MSD, middle and the LSD digits of the display.
        '
        DISPLAY_TEMPERATURE:
                new_data=2 * new_data
                first = new_data / 100
                temp = new_data // 100
                second = temp / 10
                third = temp // 10

                first = first + $E0
                POKE PORTB,first
                latch_msd = 0
                latch_msd = 1

                second = second + $E0
                POKE PORTB,second
                latch_middle = 0
                latch_middle = 1

                third = third + $E0
                POKE PORTB,third
                latch_lsd = 0
                latch_lsd = 1

                RETURN


        END
```

Fig. 7.7 (Continued)

The program branches to label *wait* unless the conversion is complete. When the conversion is complete, the converted data is stored in variable *new_data* by using the command:

    PEEK ADRES,new_data

Subroutine DISPLAY_TEMPERATURE is then called to display the temperature. The A/D converters on the PIC16C71 are 8 bits wide. An 8-bit converter has 256 possible combinations (0 to 255) of output bit patterns. With a full-scale voltage of +5 V, the accuracy of the converter is 5/256 = 19.53 mV. For example, a digital output pattern of "00010000" (i.e. decimal 16) corresponds to 312.48 mV. Similarly, a digital output pattern of "10100000" (i.e. decimal 160) corresponds to 3124.8 mV or 3.124 V and so on. As a result of this, we have to multiply the converted data by 19.53 in order to obtain the actual true voltage output of the sensor in mV. Since the output of LM35DZ is 10 mV/°C, we have to divide the output data by 10 in order to obtain the temperature directly in °C. Thus, as a summary, the digital data obtained from the sensor should be multiplied by 1.953 to give the true temperature in °C. In the program listing in Fig. 7.7, the converted data (variable *new_data*) is multiplied by 2 instead of 1.953 for simplicity since it is not easy to perform floating point operations in PIC BASIC.

Variables *first, second*, and *third* store the display data as described before.

Components Required

In addition to the standard minimum microcontroller components, the following will be required for this project:

    TIL311   Alphanumeric displays (3 off)

    LM35DZ   Digital temperature sensor IC

PROJECT 24 – Using Analogue Temperature Sensor IC with Serial A/D Converter

Function

This project shows how we can interface an analogue sensor to a PIC microcontroller using a serial A/D converter. In this project, a LM35DZ type analogue temperature sensor IC is used. The sensor is connected to a PIC16F84 type microcontroller via an ADC0831 type serial A/D converter. Figure 7.8 shows a block diagram of this project.

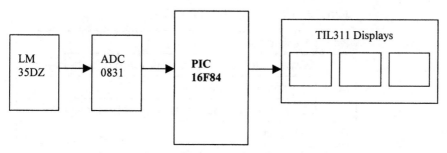

Fig. 7.8 Block diagram of Project 24

Circuit Diagram

Before looking at the circuit diagram of this project, it will be useful if we look at the ways an A/D converter can be connected to a PIC microcontroller. There are many types of A/D converters available in the market. Some converters provide serial output data such that the output is obtained from the converter each time a clock pulse is sent to the converter. These converters are very slow and are generally used where the speed of conversion is not very critical and where space saving is required. Serial A/D converters interface to a microcontroller by using only a few pins.

Standard parallel A/D converters are generally used in medium and high-speed applications. These converters are interfaced to microcontrollers using eight data pins. In addition, several control and status pins are used.

In this project, an ADC0831 type serial A/D converter is used. As shown in Fig. 7.9, ADC0831 is an 8-pin device. Table 7.3 shows the pin descriptions of this converter. The device basically interfaces to a microcontroller via 2 pins: the CLK and the DO. CLK is the clock pin and data is output from the DO pin when the CLK pin is pulsed. CS is the chip-select input and this pin should be LOW to start a conversion for proper operation of the device. Vref is the A/D converter reference pin and it

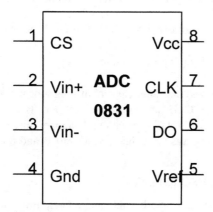

Fig. 7.9 ADC0831 pin configuration

Table 7.3 Pin descriptions of ADC0831

Pin	Description
CS	Chip select. Should be LOW to enable the device
Vin+	Analogue input (0–5 V)
Vin–	Low end analogue input. Connect to ground
Gnd	Ground
Vref	Reference voltage. Connect to +5 V
DO	Data out (when clocked)
CLK	Clock input
Vcc	+5 V power connection

should be connected to +5 V supply. Vin is where the analogue input signal should be connected.

The first clock pulse to the ADC0831 outputs a LOW bit which can be used for synchronization purposes. Actual data is then output, one bit at a time, by sending eight clock pulses to the device. This data is normally shifted into a microcontroller register so that it is available for use. The steps required to perform a conversion are as follows:

● CS is HIGH

● Clear CS to LOW to start conversion

● Send a CLK pulse to extract the sync bit (LOW)

● Send eight clock pulses to extract the 8 data bits

● Shift the bits into a register after each clock pulse

● Set CS HIGH to stop conversion

Figure 7.10 shows the complete circuit diagram of Project 23. The CLK and DO pins of the ADC0831 are connected to port B of the microcontroller. Vin is connected to an LM35DZ type analogue temperature sensor IC. Three TIL311 type alphanumeric displays are connected to port B and port A of the microcontroller.

Program Description

The program basically reads analogue temperature data from the LM35DZ sensor converts this data into digital form using the ADC0831 A/D converter and then

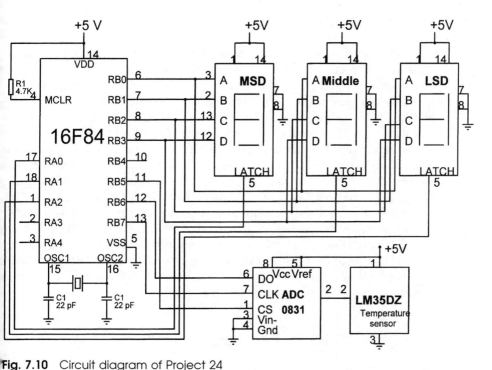

Fig. 7.10 Circuit diagram of Project 24

displays the temperature on three TIL311 type displays. The following PDL describes the operation of the program:

START

    Configure I/O ports

    DO FOREVER

        Start A/D conversion

        Send CLK pulses

        Read DO data bits

        Shift data into a register

        Stop A/D conversion

        Scale and display the temperature

    ENDDO

    END

Program Listing

The full program listing is given in Fig. 7.11. Port A is configured as outputs. Pin 6 of port B is configured as an input port and the other pins are configured as outputs. ADC0831 pins CLK and DO are assigned to 7 and 6 respectively so that they can be used to access port pins RB7 and RB6. Similarly, the CD is assigned to 5. Variables *first*, *second*, and *third* store the digits to be displayed and they have been assigned to memory locations B0–B2. Variable *new_data* stores the converted digital data. The program loop starts with label *loop* where variables *new_data* and CLK are cleared to 0. A/D conversion is then started by clearing the CS input of the ADC0831. A *FOR* loop is set up to send nine pulses to the A/D converter CLK input. Command PULSOUT sends a pulse to the CLK pin and then a new bit is read from the converter via pin 6. Variable *new_data* is shifted left and then the new bit is added to the existing data. When all 8 bits are read, variable *new_data* stores the new converted data. A/D conversion is then stopped by raising the CS pin to HIGH. Subroutine DISPLAY_TEMPERATURE is called to display the temperature on the three TIL311 displays. The data is scaled and displayed every second as in the previous project.

Using the PIC BASIC PRO Language

PIC BASIC PRO language provides a command called SHIFTIN which can be used to send out clock pulses from a specified pin and then shift data bits

```
'*****************************************************************
'
'          PROJECT:        PROJECT24
'          FILE:           PROJ24.BAS
'          DATE:           August 2000
'          PROCESSOR:      PIC16F84
'          COMPILER:       PIC BASIC
'
'
' This is a simple analogue temperature sensor project.
' An LM35DZ type analogue temperature sensor is connected
' to a serial ADC0831 type A/D converter.  CLK and DO pins
' of the converter are connected to ports RB7 and RB6 of the
' microcontroller respectively.  CS inputs is connected to
' port RB5 of the microcontroller.  In addition, 3 TIL311 type
' alphanumeric displays are connected to port B of the
' microcontroller.
'
' The program reads the analogue temperature, converts it to
' digital form and then displays on the TIL311 displays every
' second.
'
'*****************************************************************
'
' Define port addresses
symbol PORTA = 5
symbol TRISA = $85
symbol PORTB = 6
symbol TRISB = $86
'
```

Fig. 7.11 Program listing of Project 24

```
' Define ADC0831 connections
symbol CLK = 7                          'Port RB7
symbol DO = 6                           'Port RB6
symbol CS = 5                           'Port RB5
'
' Define display registers
symbol first  = B0                      'MSD digit
symbol second = B1                      'Middle digit
symbol third  = B2                      'LSD digit
'
' Define other variables
symbol new_data = B5                    'A/D data
symbol temp = B3
symbol cnt = B4
symbol secs = 1000                      'seconds delay
'
' Configure ports A and B
        POKE TRISA,0                    'Configure as outputs
        POKE TRISB,%01000000            'Configure as outputs

'
' Start of main program loop.  Send 8 pulses to the
' A/D converter and read in the serial data, 1 bit
' at a time.  Shift the data left into variable new_data
'
loop:
        new_data=0                      'Clear data to start with
        LOW CLK                         'Clear CLK line
        LOW CS                          'Start conversion
        FOR cnt = 1 TO 9                'Start pulsing CLK
                PULSOUT CLK,10
                new_data = new_data * 2
                new_data = new_data + pin6
        NEXT

        HIGH CS                         'Stop conversion
        GOSUB DISPLAY_TEMPERATURE       'Display data
        PAUSE secs                      'Delay a second
        GOTO loop                       'Go back and repeat
'
' This subroutine displays the data on 3 TIL311 type
' alphanumeric displays.  The converted data (new_data) is first
' multiplied by 2 to get the true temperature in C.
' Variables first, second, and third store the data to be
' displayed on the MSD, middle and the LSD digits of the display.
'
DISPLAY_TEMPERATURE:
        new_data=2 * new_data
        first = new_data / 100
        temp = new_data // 100
        second = temp / 10
        third = temp // 10

        POKE PORTB,first
        POKE PORTA,6
        POKE PORTA,7

        POKE PORTB,second
        POKE PORTA,5
        POKE PORTA,7

        POKE PORTB,third
        POKE PORTA,3
        POKE PORTA,7

        RETURN


END
```

Fig. 7.11 (*Continued*)

into a register through another specified pin. Figure 7.12 shows how we can program Project 24 using the PIC BASIC PRO language. The include file "MODEDEFS.BAS" is supplied with the compiler and it stores the definitions for the SHIFTIN command. Display latches *latch_msd*, *latch_middle*, and *latch_lsd* are assigned to port A pins 0, 1, and 2 respectively. ADC0831 pins CLK, DO and CS are then assigned to port B pins. Variables *first*, *second*, and *third* store the MSD, middle, and the LSD digits of the display as before. The main program starts with label loop. Here, the A/D conversion is started by lowering the CS input. The SHIFTIN command is then used to send out clock pulses and then read serial data. MSBPOST is defined in file "MODEDEFS.BAS" and this tells the SHIFTIN command to read the data after the clock is sent out. Data bits are read and shifted into variable *new_data*. Note that nine clock pulses are sent out: one for the sync bit, and eight for the actual data. The program then calls subroutine DISPLAY_TEMPERATURE to display the temperature on the three TIL311 displays. After a second delay, the program jumps to label *loop* to repeat the process.

Components Required

In addition to the standard minimum microcontroller components, the following will be required for this project:

TIL311   Alphanumeric displays (3 off)

LM35DZ   Digital temperature sensor IC

ADC0831   Analogue-to-digital converter IC

```
'****************************************************************
'
'         PROJECT:        PROJECT24
'         FILE:           PROJ24-1.BAS
'         DATE:           August 2000
'         PROCESSOR:      PIC16F84
'         COMPILER:       PIC BASIC PRO
'
'
' This is a simple analogue temperature sensor project.
' An LM35DZ type analogue temperature sensor is connected
' to a serial ADC0831 type A/D converter.  CLK and DO pins
' of the converter are connected to ports RB7 and RB6 of the
' microcontroller respectively.  The CS input is connected
' to port RB5.  In addition, 3 TIL311 type alphanumeric
' displays are connected to port B of the microcontroller.
'
' The program reads the analogue temperature, converts it to
' digital form and then displays on the TIL311 displays every
' second.
'
' This version of the program compiles only under PIC BASIC PRO
'
'****************************************************************
```

Fig. 7.12 PIC BASIC PRO listing of Project 24

```
'
Include "MODEDEFS.BAS"
'
'Define the display latches
latch_msd     var PortA.0          'MSD digit latch
latch_middle var PortA.1           'Middle digit latch
latch_lsd     var PortA.2          'LSD digit latch
'
' Define ADC0831 connections
CLK var PORTB.7                     'Port RB7
DO  var PORTB.6                     'Port RB6
CS  VAR PORTB.5                     'Port RB5
'
' Define display registers
first  var byte                    'MSD digit
second var byte                    'Middle digit
third  var byte                    'LSD digit
'
' Define other variables
new_data var byte                  'A/D data
temp     var byte
secs     con 1000                  'Seconds delay
'
' Start of main program loop.  Use command SHIFTIN
' to send clock pulses to the A/D converter and then
' read the converted bits into variable new_data.
'
        TRISA = 0
        TRISB = %01000000

loop:
        CS = 0                          'Start conversion
        SHIFTIN DO,CLK,MSBPOST,[new_data\9]
        CS = 1                          'Stop conversion
        GOSUB DISPLAY_TEMPERATURE       'Display data
        PAUSE secs                      'Delay a second
        GOTO loop                       'Go back and repeat


'
' This subroutine displays the data on 3 TIL311 type
' alphanumeric displays.  The converted data (new_data) is first
' multiplied by 2 to get the true temperature in C.
' Variables first, second, and third store the data to be
' displayed on the MSD, middle and the LSD digits of the display.
'
DISPLAY_TEMPERATURE:
        new_data=2 * new_data
        first = new_data / 100
        temp = new_data // 100
        second = temp / 10
        third = temp // 10

        PORTB = first
        latch_msd = 0
        latch_msd = 1

        PORTB = second
        latch_middle = 0
        latch_middle = 1

        PORTB = third
        latch_lsd = 0
        latch_lsd = 1

        RETURN


END
```

Fig. 7.12 (Continued)

PROJECT 25 – Using Analogue Temperature Sensor IC with Parallel A/D Converter and LCD

Function

This is a larger project which shows how we can connect a parallel A/D converter and an LCD display to a PIC microcontroller. The project reads the analogue temperature from an LM35DZ type sensor, converts it to digital form and then displays the temperature on an LCD display. Because of the need for a large number of I/O ports, a PIC16C642 type microcontroller is used in this project.

Circuit Diagram

The block diagram of this project is shown in Fig. 7.13. As shown in the circuit diagram in Fig. 7.14, the output of an LM35DZ type analogue temperature sensor is connected to an ADC0804 type 8-bit parallel A/D converter. The output pins of this converter are connected to port C of a PIC16C642 type microcontroller. The microcontroller drives a standard LCD display via its port A pins.

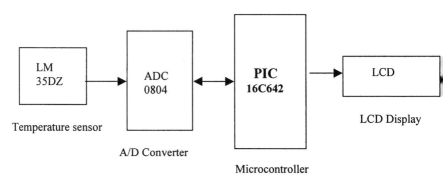

Fig. 7.13 Block diagram of Project 25

PIC16C642 is a 28-pin microcontroller. This is a fast microcontroller which can operate at up to 20 MHz. The device features 22 I/O pins with high current sink/ source capabilities. Two analogue comparators are provided with a programmable on-chip voltage reference. In addition, the microcontroller provides an 8-bit timer/ counter with 8-bit programmable prescaler, a power-up timer, watchdog timer, power saving sleep mode, and program memory parity error checking circuitry. The PIC16C642 microcontroller is erased under ultraviolet light and the device is programmed via two pins, using a serial programming algorithm. The I/O ports are named as port A (or RA0–RA5), port B (or RB0–RB7), and port C (or RC0–RC7) Port A is 6 bits wide while the other two ports are 16 bits each.

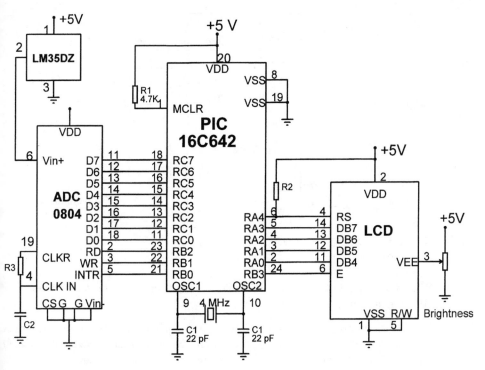

Fig. 7.14 Circuit diagram of Project 25

ADC0804 is an 8-bit parallel A/D converter, manufactured by the National Semiconductor Corporation. This is a 20-pin device with a conversion time of 100 microseconds. As shown in Fig. 7.15, these converters interface to a microcontroller using the following pins (only the pins used in a standard application are shown):

DB0–DB7	8 data output pins
RD	Read input
WR	Write input
INTR	Interrupt output
CLKR/CLKIN	Clock control inputs
Vin +	Positive analogue input

DB0 to DB7 are the digital output lines and the converted data appears on these eight lines. An 8-bit converter has 256 possible combinations (0 to 255) of output bit patterns. With a full-scale voltage of +5 V, the accuracy of the converter is 5/256 = 19.53 mV.

RD is the read data control pin and when RD is LOW, output data appears on the 8 output pins. When RD is HIGH, the output is not available.

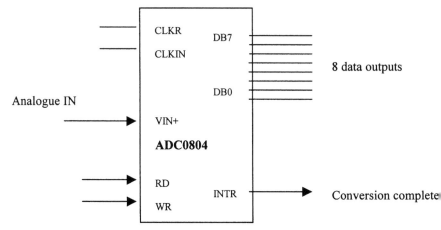

Fig. 7.15   ADC0804 Functional pin configuration

WR input is normally at logic HIGH and this input should be set to LOW and then HIGH again for a conversion to start.

INTR is the interrupt output of the A/D converter. A HIGH to LOW pulse is generated on this pin when a conversion is complete. This output is usually used to generate an interrupt in the microcontroller so that the converted data can be read.

ADC0804 contains an internal oscillator and it is required to connect an external resistor and capacitor to pins CLKR and CLKIN to start this oscillator.

Vin+ is the pin where the analogue input voltage should be applied.

To make a single conversion the operation of the A/D converter can be summarized using the following steps:

- Set WR and RD HIGH

- Start conversion by setting WR LOW

- Set WR back to HIGH

- Detect end of conversion when INTR goes LOW (usually by interrupt)

- Set RD LOW and read data from DB0 to DB7

- Set RD HIGH

The above process is of course repeated when more than one conversion is required. In Fig. 7.14, the output of the sensor (pin 2) is connected to the Vin+ pin (pin 6) of the ADC0804. Port C pins of the microcontroller are connected to the DB pins of the A/D converter. Pins RD, WR and INTR are connected to RB2, RB1 and RB0 respectively. A 10K resistor and a 150 pF capacitor are connected to the clock inputs CLKR and CLKIN to start the internal oscillator.

The LCD is connected in the standard way, i.e. port pins RA0–RA3 are connected to the data pins of the LCD. Pin RA4 is connected to the RS pin and pin RB3 is connected to the E pin of the LCD.

Program Description

The program is based on the ADC0804 generating an external interrupt to the microcontroller. Upon receipt of this interrupt, the microcontroller reads the converted data and displays it on the LCD. The following PDL describes the operation of the program:

START

    Configure microcontroller I/O ports

    Call subroutine INITIALIZE to initialize the A/D and microcontroller

    Start A/D conversion

    DO FOREVER

        Wait for an interrupt

    ENDDO

END

Subroutine INITIALIZE

START

    SET RD and WR pins HIGH

    Configure the microcontroller to receive external interrupts

END

Interrupt Service Routine

START

    Read converted data from the A/D

    Scale the result

    Display the result on the LCD

    Delay a second

    Re-start A/D conversion

END

Program Listing

The full program listing is shown in Fig. 7.16. The program has been developed using the PIC BASIC PRO language. At the beginning of the program A/D pins RD and WR are assigned to port B pins 2 and 1 respectively.

Microcontroller register CMCON is set to 7 (i.e. bit pattern "00000111"). When power is applied to the PIC16C642 microcontroller, the RA0–RA3 are in analogue comparator mode. Setting register CMCON to 7 (see the PIC16C642 data sheet) sets internal bits CM0–CM2 to 1s and this configures the RA0–RA3 ports as digital I/O ports.

Port B bit 0 is then configured as an input and the other bits of this port are configured as digital outputs. Port C is configured as an 8-bit input port since this port receives the converted data. Port A is set as a digital output as it drives the LCD display.

The interrupt service routine is declared as the label ISR. The program then calls subroutine INTIALIZE which initializes the A/D converter and the microcontroller. RD and WD inputs of the A/D converter are set HIGH so that the chip is ready for conversion.

```
'****************************************************************
'
'           PROJECT:        PROJECT25
'           FILE:           PROJ25.BAS
'           DATE:           August 2000
'           PROCESSOR:      PIC16C642
'           COMPILER:       PIC BASIC PRO
'
'
' This is a simple LCD based analogue temperature sensor project.
' An LM35DZ type analogue temperature sensor is connected
' to a parallel ADC0804 type A/D converter.  RD and WR pins
' of the converter are connected to pins RB2 and RB1 of the
' microcontroller respectively.  Pin INTR of the A/D converter
' is connected to the external interrupt pin (RB0) of the
' microcontroler.
'
' The program reads analogue temperature from the LM35DZ via
' the A/D converter.  The temperature is then converted into
' digital form and is displayed on a standard LCD.  This
' process is repeated every second.
'
' This version of the program compiles only under PIC BASIC PRO
'
'****************************************************************
'
'
'Define the ADC0804 pins
RD var PortB.2
WR var PortB.1
'
' Define other symbols and variables
new_value var byte
i var byte
secs con 1000                           'Seconds delay
```

Fig. 7.16 Program listing of Project 25

```
'
'
            CMCON = 7                       'Set port A as digital I/O
            TRISB = %00000001               'Configure port B
            TRISC = %11111111               'Configure port C
            TRISA = 0                       'Configure port A

            PAUSE secs                      'Wait a while for LCD startup
            LCDOUT $FE, 1, "Waiting..."     'Send a message to LCD

            ON INTERRUPT GOTO ISR           'Define ISR routine
            GOSUB INITIALIZE                'Initialize
' Start conversion
            WR = 0                          'Pulse WR pin
            WR = 1                          '
'
' Start of main program where we wait here for an interrupt.
' When the A/D conversion is complete, an external interrupt
' is generated on pin RB0 and the program automatically jumps
' to label ISR where the converted temperature is read, scaled
' and displayed on the LCD.
'
loop:       I = I + 1                       'Dummy statement
            GOTO loop                       'Wait for interrupr


'
' This subroutine intializes the A/D converter and the
' microcontroller.  Pins RD and WR of teh A/D converter
' are set HIGH so that they are ready for conversion.
' The OPTION register of the PIC16C642 microcontroller is
' configured so that external interrupts can be recognized
' on the falling edge of input RB0.  Also, the INTCON register
' is configured so that global as well as external interrupts
' on RB0 are enabled.
'
INITIALIZE:
            RD = 1                          'Set A/D RD to 1
            WR = 1                          'Set A/D WR to 1
            OPTION_REG = 0                  'Set interrupt on falling edge of RB0
            INTCON = $90                    'Enable global interrupts,
                                            'Enable RB0 external interrupts

            RETURN


'
' This is the interrupt service routine.  The program jumps here
' whenever the A/D conversion is complete.  Here, the converted
' data is read, scaled and then output to the LCD.  The ISR also
' re-starts the A/D conversion for the next sample.
'
DISABLE
ISR:
            RD = 0                          'Enabe reading
            new_value = PORTC               'Read converted value
            new_value = 2 * new_value
            RD = 1                          'Disable reading
            LCDOUT $FE, 1,"TEMP = ",dec2 new_value
            PAUSE secs                      'Delay a second
            INTCON.1 = 0                    'Clear interrupt flag
            WR = 0                          'Re-start conversion
            WR = 1                          '
            RESUME                          'Back to where interrupted
ENABLE


END
```

Fig. 7.16 (*Continued*)

External interrupts are configured by using the microcontroller registers OPTION_ REG and INTCON. Bit 6 of the microcontroller OPTION register (see the data sheet) is cleared to 0 so that interrupts can be accepted on the falling edge of the RB0 input. Since the other bits of this register are not relevant, the complete register is cleared to 0. INTCON register (see PIC16C642 data sheet) is set to the bit pattern "10010000" (i.e. hexadecimal value $90) so that external interrupts on pin RB0 can be accepted by the microcontroller.

The program then enters a loop and waits for an interrupt from the A/D converter. Note that when using the PIC BASIC PRO language, interrupts are checked between the statements and if we use the statement:

    loop: GOTO loop

then interrupts may not be recognized by the microcontroller. As a result of this a *dummy* statement is used ($I = I + 1$) inside the loop while waiting for an interrupt.

The interrupt service routine is entered by label ISR. Just before entering this routine further interrupts are disabled. The ISR routine clears pin RD of the A/D converter and then reads the converted value into variable *new_value*. This value is then scaled by multiplying by 2 as described in the earlier A/D converter-based projects. The result is displayed on LCD in the format:

    TEMP = xx

Where *xx* is the 2-digit temperature. The program then delays for a second. The interrupt flag (bit 1 of the INTCON register) is cleared so that further interrupts can be accepted by the microcontroller. Pin WR of the A/D is pulsed to start a new conversion and the program is returned from the interrupt service routine by executing the RESUME statement.

Components Required

The following components are required for this project:

C1	22 pF capacitors (2 off)
R1	4.7K resistor
R2, R3	10K resistors
C2	150 pF capacitor
	PIC16C642 microcontroller
	LM35DZ temperature sensor IC
	ADC0804 A/D converter IC
	LCD

Chapter 8

RS232 Serial Communication Projects

RS232 is a serial communications standard which enables data to be transferred in serial form between two devices. Data is transmitted and received in serial "bit stream" from one point to another. Standard RS232 is suitable for data transfer to about 50 m, although special low-loss cables can be used for extended distance operation. Four parameters specify an RS232 link between two devices. These are: *baud rate, data width, parity*, and the *stop bits*, as described below.

Baud rate: The baud rate (bits per second) determines how much information is transferred over a given time interval. A baud rate can usually be selected between 110 and 76 800 baud, e.g. a baud rate of 9600 corresponds to 9600 bits per second.

Data width: The data width can either be 8 bits or 7 bits depending upon the nature of the data being transferred.

Parity: The parity bit is used to check the correctness of the transmitted or received data. Parity can either be even, odd, or no parity bit can be specified at all.

Stop bit: The stop bit is used as the terminator bit and it is possible to specify either 1 or 2 stop bits.

Serial data is transmitted and received in frames where a frame consists of:

- 1 start bit
- 7 or 8 data bits
- optional parity bit
- 1 stop bit

In many applications 10 bits are used to specify an RS232 frame, consisting of 1 start bit, 8 data bits, no parity bit and 1 stop bit. For example, character "A" has the ASCII bit pattern "01000001" and is transmitted as shown in Fig. 8.1 with 1 start bit, 8 data bits, no parity and 1 stop bit. The data is transmitted least significant bit first.

When 10 bits are used to specify the frame length, the time taken to transmit or receive each bit can be found from the baud rate used. Table 8.1 gives the time taken for each bit to be transmitted or received for most commonly used baud rates.

START   1   0   0   0   0   0   1   0   STOP

Fig. 8.1 Transmitting character "A" (bit pattern 01000001)

Table 8.1 Bit times for different baud rates

Baud Rate	Bit Time
300	3.33 ms
600	1.66 ms
1200	833 μs
2400	416 μs
4800	208 μs
9600	104 μs
19200	52 μs

8.1 RS232 Connectors

As shown in Fig. 8.2, two types of connectors are used for RS232 communications. These are the 25-way D-type connector (known as DB25) and the 9-pin D-type connector (known as DB9). Table 8.2 lists the most commonly used signal names for both DB9 and DB25 type connectors. The used signals are:

SG: Signal ground. This pin is used in all RS232 cables.

RD: Received data. Data is received at this pin. This pin is used in all two-way communications.

TD: Transmit data. Data is sent out from this pin. This pin is used in all two-way communications.

RTS: Request to send. This signal is asserted when the device requests data to be sent.

CONN-D25

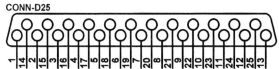

CONN-D9

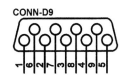

Fig. 8.2 RS232 connectors

Table 8.2 Commonly used RS232 signals

Description	Signal	9-pin	25-pin
Carrier Detect	CD	1	8
Receive Data	RD	2	3
Transmit Data	TD	3	2
Data Terminal Ready	DTR	4	20
Signal Ground	SG	5	7
Data Set Ready	DSR	6	6
Request to Send	RTS	7	4
Clear to Send	CTS	8	5
Ring Indicator	RI	9	22

CTS: Clear to send. This signal is asserted when the device is ready to accept data.

DTR: Data terminal ready. This signal is asserted to indicate that the device is ready.

DSR: Data set ready. This signal tells the device at the other end to indicate that it is ready.

CD: Carrier detect. This signal indicates that a carrier signal has been detected by a modem connected to the line.

In some RS232 applications it is sufficient to use only the pins SG, RD and TD. Also, in some applications (e.g. when two similar devices are connected together) it is necessary to twist pins RD and TD so that the transmit pin of one device is connected to the receive pin of the other device and vice versa.

8.2 RS232 Signal Levels

RS232 is bi-polar and a voltage +3 to +12 volts indicates an ON state (or SPACE), while −3 to −12 volts indicate an OFF state (or MARK). In practice, the ON and OFF states can be achieved with lower voltages.

Standard TTL logic devices (including the PIC microcontrollers) operate with TTL compatible logic levels between the voltages of 0 and +5 volts. Voltage level converter ICs are used to convert between the TTL and RS232 voltage levels. One such popular IC is the MAX232, manufactured by MAXIM, and operates with a +5 volt supply. This is a 16-pin DIL chip incorporating two receivers and two transmitters (see Fig.

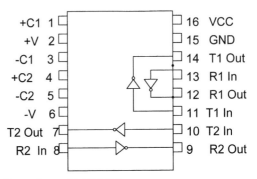

Fig. 8.3   Pin configuration of MAX232

8.3) and the device requires four external capacitors for proper operation. RS232 voltages are inverted with respect to normal logic. That is, we normally represent a logic "1" using the higher voltage (e.g +5 V). In RS232 it is the negative logic: a "1" is the lower voltage (negative) and a "0" is the higher voltage.

8.3 Controlling the RS232 Port

PIC microcontrollers can be programmed to input and output asynchronous serial data from any of their port pins. The SEROUT command in PIC BASIC and PIC BASIC PRO languages allows the programmer to output serial data easily. The format of this command is:

SEROUT Pin, Mode, Item, . . .

This command sends asynchronous serial data in 8-bit data format with no parity and 1 stop bit to the chosen pin. As shown in Table 8.3, *mode* defines the baud rate and the type of output required. In normal RS232 type connections with level converters, normal TTL output (e.g. T2400) should be chosen. In most cases we can connect the output pin of a PIC microcontroller directly to an RS232 compatible receiver. In such cases, the inverted output (e.g. N2400) should be used for proper operation. Using the SEROUT command we can send a string constant, a numeric value or the ASCII representation of a decimal value.

Similarly, the SERIN command in PIC BASIC and PIC BASIC PRO languages enables the programmer to input asynchronous serial data. The format of this command is:

SERIN Pin,Mode,(Qual),Item

This command receives serial asynchronous data in 8-bit data format with no parity and 1 stop bit. The *mode* defines the baud rate and the type of input required as shown in Table 8.4. *Qual* is a qualifier and it can be a constant, a variable, or a string

Table 8.3. SEROUT command mode values

Symbol	Baud Rate	Mode
T300	300	TTL true
T1200	1200	TTL true
T2400	2400	TTL true
T9600	9600	TTL true
N300	300	TTL inverted
N1200	1200	TTL inverted
N2400	2400	TTL inverted
N9600	9600	TTL inverted
OT300	300	Open drain
OT1200	1200	Open drain
OT2400	2400	Open drain
OT9600	9600	Open drain
ON300	300	Open source
ON1200	1200	Open source
ON2400	2400	Open source
ON9600	9600	Open source

Table 8.4 SERIN command mode values

Symbol	Baud Rate	Mode
T300	300	TTL true
T1200	1200	TTL true
T2400	2400	TTL true
T9600	9600	TTL true
N300	300	TTL inverted
N1200	1200	TTL inverted
N2400	2400	TTL inverted
N9600	9600	TTL inverted

constant. We can define more than one qualifier and also more than one item. It is important to note that the data received should match the qualifiers.

PIC BASIC PRO language also supports HSEROUT and SEROUT2 commands for serial data communication with other qualifiers. These commands are outside the scope of this book and the reader can obtain more information on these commands by referring to the relevant programming guide.

LET BASIC language provides the RSOUT command for sending out serial data. The format of this command is:

RSOUT value

This command outputs the *value* using the serial asynchronous communications protocol. The pin used must be defined using the INIT command before using the RSOUT:

INIT rsin pin,rsout pin,dtr

Where *rsin pin* is the port pin used for serial input, *rsout pin* is the port pin used for serial data output and *dtr* is the optional *data terminal ready* signal.

Similarly, the RSIN command in LET BASIC is used to receive serial data. The format of this command is:

variable = RSIN

where the *variable* stores the serial data received. In LET BASIC, the data format is 8-bit data, no parity and 1 stop bit. The baud rate is not programmable and it depends upon the crystal frequency used. With a 4 MHz crystal, the baud rate is 9600. Similarly, with an 8 MHz crystal, the baud rate is 19 200.

PROJECT 26 – Output a Simple Text Message from an RS232 Port

Function

This project shows how we can interface a PIC microcontroller to an external RS232 compatible device (e.g. an RS232 visual display unit, or COM1 or COM2 port of a PC) and send a text message to this device. The text message "PIC TEST" is sent out continuously from the microcontroller. The frame format used in this project is 2400 baud, 8 data bits, no parity, and 1 stop bit.

Circuit Diagram

The block diagram of Project 26 is shown in Fig. 8.4. The complete circuit diagram is shown in Fig. 8.5 where a PIC16F84 type microcontroller is used in this project. Pin (

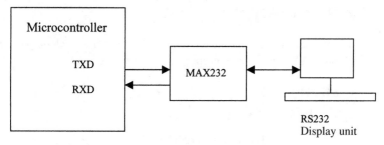

Fig. 8.4 Block diagram of Project 26

of port B (RB0) is assumed to be the serial output of the microcontroller, and is connected to pin 10 of the MAX232 voltage converter IC. The output of this IC (pin 7) can be connected to the input of a COM1 (or COM2) port of a PC, or to the input of an RS232 visual display unit. Similarly, the output of the external RS232 device is connected to bit 1 of port B (RB1) via the MAX232 IC. Thus, RB1 is assumed to be the serial input of the microcontroller. A terminal emulation software can be activated on the PC to receive and display any data arriving at its serial port.

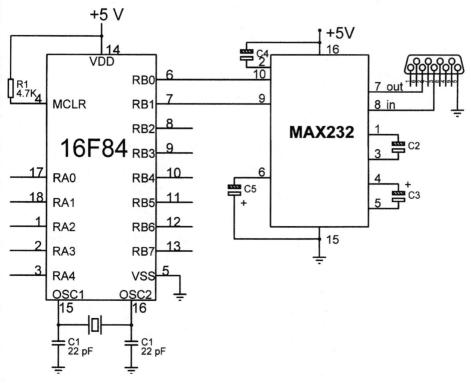

Fig. 8.5 Circuit diagram of Project 26

Program Description

The program is very simple as it uses the built-in serial output command. The following PDL describes the operation of the program:

START

    DO FOREVER

        Send text "PIC TEST" to the chosen RS232 port

    ENDDO

END

Program Listing

The program listing is given in Fig. 8.6. Symbol *sout* is assigned to 0 and this variable is used to define the serial output port. Symbols *cr* and *lf* are assigned to the carriage-return and line-feed characters respectively. The program starts with label *loop* where the SEROUT command is used to send data to the serial port:

SEROUT sout,T2400,("PIC TEST",cr,lf)

```
'****************************************************************
'
'          PROJECT:        PROJECT26
'          FILE:           PROJ26.BAS
'          DATE:           August 2000
'          PROCESSOR:      PIC16F84
'          COMPILER:       PIC BASIC
'
'
' This project sends the text message "PIC TEST" every second
' to an RS-232 compatible serial device.  Carriage-return and
' line-feed characters are sent at the end of the message.
' A MAX232 type RS-232 level converter IC is used to translate
' between TTL and RS-232 signals.  Pin 0 of port B (RB0) is
' used as the output pin.
'
'****************************************************************
symbol sout = 0                    'Pin 0 is used as output
symbol cr   = 13                   'Carriage-return character
symbol lf   = 10                   'Line-feed character
symbol sec  = 1000                 'Second delay

'
' Use SEROUT command to send out message PIC TEST together
' with carriage-return and line-feed.
'
loop:
        SEROUT sout,T2400,("PIC TEST",cr,lf)
        PAUSE sec
        GOTO loop

        END
```

Fig. 8.6 PIC BASIC program listing of Project 26

Where the baud rate is chosen as 2400 and the message is terminated with a carriage-return and line-feed pair. The following message is sent to the external RS232 unit:

PIC TEST

PIC TEST

PIC TEST

....

....

The data output is repeated with a 1 second delay between each output sample.

Connecting the Output Directly

It is possible to connect the output pin of a PIC microcontroller directly to an RS232 receiver unit without using a voltage level converter. When this is done, we have to invert the output pin by preceding the baud rate with character "N" as shown below:

SEROUT sout,N2400,("PIC TEST",cr,lf)

Using the PIC BASIC PRO Language

Serial port programming in PIC BASIC PRO is similar to the listing given in Fig. 8.6. The main difference is that square brackets are used in PIC BASIC PRO as shown below:

SEROUT sout,T2400,["PIC TEST",cr,lf]

Using the LET BASIC Language

Programming the serial port is different with LET BASIC. The baud rate is not programmable and depends upon the crystal frequency chosen. With a 4 MHz crystal, the baud rate is 9600. Similarly, with an 8 MHz crystal, the baud rate is 19 200. Also, LET BASIC assumes that the output pin is inverted and thus there is no need to use a voltage level converter IC.

Figure 8.7 shows a LET BASIC program which sends the numbers 0–9, followed by the characters A–Z to an external RS232 unit. When the program is run, the following data is sent out:

0123456789

ABCDEFGHIJKLMNOPQRSTUVWXYZ

0123456789

....

....

```
REM*****************************************************************
REM
REM         PROJECT:          PROJECT26
REM         FILE:             PROJ26-1.BAS
REM         DATE:             August 2000
REM         PROCESSOR:        PIC16F84
REM         COMPILER:         LET BASIC
REM
REM
REM This project sends the text message "PIC TEST" every
REM second to an RS-232 compatible serial device.  Carriage-
REM return and line-feed characters are sent at the end of the
REM message.
REM
REM A MAX232 type RS-232 level converter IC is used to translate
REM between TTL and RS-232 signals.  Pin 0 of port B (RB0) is
REM used as the output pin.
REM
REM*****************************************************************

DEVICE 16F84
INCLUDE serial
DEFINE portB=00000010
DIM i
symbol sout=B.0
symbol sin=B.1
INIT serial sin,sout

loop:
        FOR i = 48 TO 57
        RSOUT(i)
        NEXT i

        RSOUT(13)
        RSOUT(10)

        FOR i = 65 TO 90
        RSOUT(i)
        NEXT i

        RSOUT(13)
        RSOUT(10)

        DELAYMS(200)
        GOTO loop

        END
```

Fig. 8.7 LET BASIC program listing

The device type is defined at the beginning of the program. Pin 0 of port B is then defined as the output pin and symbol *sout* is assigned to this pin. Similarly, pin 1 is defined as an input although this pin is not used in this project. The command:

INIT serial sin,sout

defines that we are using a serial port and that the port output pin is *sout* and the port input pin is *sin*. The program then starts with label *loop* where a *FOR* loop is used to send all the numbers 0 to 9 to the serial port using the commands:

FOR i = 48 TO 57

     RSOUT(i)

NEXT i

After sending a carriage-return, line-feed pair the program sends out all the letters from A to Z:

FOR i = 65 TO 90

      RSOUT(i)

NEXT i

Carriage-return and line-feed characters are sent out again and the program repeats by jumping to label loop.

Components Required

The following components are required in addition to the standard microcontroller components:

C2, C3, C4, C5   22 μF 25 V electrolytic capacitors

                9-way RS232 socket

                MAX232 IC

PROJECT 27 – RS232 Input/Output Example

Function

This project shows how we can input and output RS232 compatible serial data using a PIC microcontroller. In this project, the message "Enter a character:" is displayed on user's device. After accepting the user's character, the program displays the message "You have typed:" and displays the character entered by the user. This process is repeated forever.

Circuit Diagram

The circuit diagram of this project is the same as in Fig. 8.5.

Program Description

The program uses the built-in serial input/output commands of the PIC BASIC language. The following PDL describes the operation of the program:

START

    DO FOREVER

        Send message "Enter a character" on user's RS232 device

        Get a character from user's device

Send message "You have typed:" on user's device, followed by the character typed by the user

Delay a second

ENDDO

END

Program Listing

The PIC BASIC program listing of this project is shown in Fig. 8.8. Variables *sout* and *sin* are assigned to the output and input pins of port B. The SEROUT command is used to send out the message "Enter a character:" in 2400 baud rate. The SERIN command then waits until the user enters a character and the received character is stored in variable B0. The SEROUT command then displays the message "You have typed:", followed by the character typed by the user. A carriage-return and line-feed character pair are then sent to the device so that the cursor moves to the beginning

```
'*****************************************************************
'
'          PROJECT:        PROJECT27
'          FILE:           PROJ27.BAS
'          DATE:           August 2000
'          PROCESSOR:      PIC16F84
'          COMPILER:       PIC BASIC
'
'
' This project shows how we can input and output RS-232
' compatible serial data from the microcontroller.
'
' The program outputs the text message "Enter a character:"
' and then waits for the user to enter a character.  The
' character entered by the user is received and displayed
' by the microcontroller.
'
' A MAX232 type RS-232 level converter IC is used to translate
' between TTL and RS-232 signals.  Pin 0 of port B (RB0) is
' used as the output pin. Pin 1 of port B (RB1) is used as the
' input pin.
'
'*****************************************************************
symbol sout = 0                'Pin 0 is used as output
symbol sin  = 1                'Pin 1 is used as input
symbol cr   = 13               'Carriage-return character
symbol lf   = 10               'Line-feed character
symbol sec  = 1000             'Second delay

'
' Use SEROUT command to send out message and use SERIN to read
' message from the serial port.
'
loop:
        SEROUT sout,T2400,("Enter a character:")
        SERIN sin,T2400,B0
        SEROUT sout,T2400,("   You have typed:",B0,13,10)
        PAUSE sec
        GOTO loop

        END
```

Fig. 8.8 Program listing of Project 27

of the next line. The program repeats after a second delay. The following is an example display on the user's terminal:

Enter a character:d   You have typed:d

Enter c character:t   You have typed:t

....

....

PROJECT 28 – A Simple Calculator Using an RS232 Port

Function

This is a simple calculator project based upon the PIC16F84 type microcontroller. The microcontroller is connected to an RS232 serial terminal. The user can perform simple addition, subtraction, multiplication, and division of integer numbers using the microcontroller.

Circuit Diagram

The circuit diagram of this project is the same as in Fig. 8.5, i.e. pins 0 and 1 of port B are connected to an external RS232 device via a MAX232 type voltage level converter IC.

Program Description

The program operates as a simple integer calculator. When power is applied to the microcontroller, a menu is displayed on the user's terminal and the user is prompted to enter two numbers and the operation to be performed. The result of the operation is then displayed on user's terminal and the process repeats. A typical dialogue is given below:

A SIMPLE CALCULATOR PROGRAM

Enter 2 integer numbers and the operation

Valid operations are + – */

First number:3

Second number:5

Operation:+

Result = 8

A SIMPLE ...

The following PDL describes the operation of the program:

START

    DO FOREVER

        Display heading

        Get number 1

        Get number 2

        Get operation required

        IF operation = "+" **THEN**

            ADD the numbers

        ELSEIF operation = "–" **THEN**

            SUBTRACT the numbers

        ELSEIF operation = "*" **THEN**

            MULTIPLY the numbers

        ELSEIF operation = "/" **THEN**

            DIVIDE the numbers

        ENDIF

        Display the result

    ENDDO

END

Program Listing

The program listing of Project 28 is shown in Fig. 8.9. Symbols sout and sin are assigned to 0 and 1 to represent pins 0 and 1 of port B respectively. Symbols cr and lf are defined as the carriage return and line feed respectively. The two numbers are stored in variables *no1* and *no2*. The operation and the result are stored in variables *oper* and *result* respectively.

Several SEROUT commands are used to display the heading and the user is prompted to enter the first number. This number is read using command SERIN and is stored in variable *no1*. Note that the "#" character precedes the variable name so that the decimal value in ASCII is converted and stored in the variable. The data input is terminated when the user presses the *enter* key. The program then displays the message "Second number;" and again waits for the user to enter the second number which is stored in variable *no2*. The operation to be performed is then displayed on the user's terminal where the user is expected to enter a valid

```
'**************************************************************
'
'        PROJECT:        PROJECT28
'        FILE:           PROJ28.BAS
'        DATE:           August 2000
'        PROCESSOR:      PIC16F84
'        COMPILER:       PIC BASIC
'
'
' This is a simple integer calculator program.  The program
' shows how we can input and output data through a serial
' port.  The program is MENU based and is simple to use.
'
' In this program the user is required to enter 2 integer numbers
' and then the required operation (+ - * /).  The program performs
' the operation and then sends the result to the configured RS-232
' port.
'
' A MAX232 type RS-232 level converter IC is used to translate
' between TTL and RS-232 signals.  Pin 0 of port B (RB0) is
' used as the output pin. Pin 1 of port B (RB1) is used as the
' input pin.
'
'**************************************************************
symbol sout = 0                    'Pin 0 is used as output
symbol sin  = 1                    'Pin 1 is used as input
symbol cr   = 13                   'Carriage-return character
symbol lf   = 10                   'Line-feed character
symbol sec  = 1000                 'Second delay
symbol no1  = B0
symbol no2  = B1
symbol oper = B2
symbol result = B3
'
' Display the calculator MENU and wait for user inputs.
'
loop:
        SEROUT sout,T2400,(cr,lf,lf,"A Simple Calculator Program",cr,lf)
        SEROUT sout,T2400,("============================",cr,lf)
        SEROUT sout,T2400,("Enter 2 integer numbers and the operation",cr,lf)
        SEROUT sout,T2400,("Valid operations are + - * /",cr,lf,lf)
   '
   ' Get the two numbers
   '
        SEROUT sout,T2400,("First number:")
        SERIN  sin,T2400,#no1
        SEROUT sout,T2400,(#no1)
        SEROUT sout,T2400,(cr,lf,"Second number:")
        SERIN  sin,T2400,#no2
        SEROUT sout,T2400,(#no2)
    '
    ' Get the required operation
    '
        SEROUT sout,T2400,(cr,lf,"Operation:")
        SERIN  sin,T2400,oper
        SEROUT sout,T2400,(oper,cr,lf)
  '
  ' Now perform the required operation
  '
        IF oper = "+" THEN add
        IF oper = "-" THEN subtract
        IF oper = "*" THEN multiply
        IF oper = "/" THEN divide
        GOTO loop


'
' Add the 2 numbers
'
```

Fig. 8.9 Program listing of Project 28

```
add:
        result = no1 + no2
        GOTO disply

'
' Subtract the 2 numbers
'
subtract:
        result = no1 - no2
        GOTO disply
'
' Multiply the 2 numbers
'
multiply:
        result = no1 * no2
        GOTO disply
'
' Divide the 2 numbers
'
divide:
        result = no1 / no2
'
' Now send the result to the RS-232 port
'
disply:
        SEROUT sout,T2400,("Result=",#result)
        GOTO loop

        END
```

Fig. 8.9   (Continued)

character. Several IF statements are used to direct the program flow to the appropriate routines so that the required operation can be performed. For example if the operation is character "+" then the program jumps to label ADD and so on The result of the operation, which is in variable *result* is stored using a SEROUT command.

Adding Password to the Calculator Program (Using the EEPROM)

This section shows how we can use the EEPROM memory which is available on the PIC16F84, 16C84, 16F87x series of microcontrollers. In this example, we store a user defined password character in the first location of the EEPROM memory and then check this password every time the program is restarted (i.e. after a power-on or after a reset).

The same circuit as in Fig. 8.5 is used here. PIC16F84 contains a 64-byte EEPROM memory, addressed from 0 to 63. PIC BASIC commands (also PIC BASIC PRO READ and WRITE are available to read and store data in the EEPROM memory LET BASIC offers commands STORE and EEDATA to write and read from the built-in EEPROM memory.

The program listing is shown in Fig. 8.10. At the beginning of the program the first location of the EEPROM (address 0) is read into variable *eeprm*. If this is the first time we are running the program then this address contains all 1s (i.e. decimal 255

```
'****************************************************************
'
'         PROJECT:        PROJECT28
'         FILE:           PROJ28-1.BAS
'         DATE:           August 2000
'         PROCESSOR:      PIC16F84
'         COMPILER:       PIC BASIC
'
'
' This is a simple integer calculator program.  The program
' shows how we can input and output data through a serial
' port.  The program is MENU based and is simple to use.
'
' This program is similar to the previous program but a password
' facility is added using the on-chip EEPROM.  A password is stored
' in the EEPROM and the program continues only if the user knows
' the correct password.
'
' In this program the user is required to enter 2 integer numbers
' and then the required operation (+ - * /).  The program performs
' the operation and then sends the result to the configured RS-232
' port.
'
' A MAX232 type RS-232 level converter IC is used to translate
' between TTL and RS-232 signals.  Pin 0 of port B (RB0) is
' used as the output pin. Pin 1 of port B (RB1) is used as the
' input pin.
'
'****************************************************************
symbol sout = 0                  'Pin 0 is used as output
symbol sin  = 1                  'Pin 1 is used as input
symbol cr   = 13                 'Carriage-return character
symbol lf   = 10                 'Line-feed character
symbol sec  = 1000               'Second delay
symbol no1  = B0                 '
symbol no2  = B1
symbol oper = B2
symbol result = B3
symbol eeprm = B4
symbol pwd = B5


' Check the first EEPROM location.  If not blank (255) then
' goto label chk_pwd to check the password, otherwise this
' is the first time and ask for a new password.
'
        READ 0, eeprm.                        'Check EEPROM address 0
        IF eeprm <> 255 THEN chk_pwd          'If blank then ask
        SEROUT sout,T2400,("New Password:")   'for a new password
        SERIN sin,T2400,pwd                   '
        WRITE 0,pwd                           'Write the new password
        GOTO loop                             'into location 0
'
' This was not the first time, check the password.
'
chk_pwd:
        SEROUT sout,T2400,("Password:")       'Ask the password
        SERIN sin,T2400,pwd                   'Get the password
        IF pwd = eeprm THEN loop              'Correct password ?
        SEROUT sout,T2400,(cr,lf,"Wrong",cr,lf)
        goto chk_pwd

'
' Display the calculator MENU and wait for user inputs.
'
loop:
        SEROUT sout,T2400,(cr,lf,lf,"A Simple Calculator Program",cr,lf)
        SEROUT sout,T2400,("=============================",cr,lf)
        SEROUT sout,T2400,("Enter 2 integer numbers and the operation",cr,lf)
        SEROUT sout,T2400,("Valid operations are + - * /",cr,lf,lf)
```

Fig. 8.10 Using the EEPROM memory

```
'
' Get the two numbers
'
        SEROUT sout,T2400,("First number:")
        SERIN sin,T2400,#no1
        SEROUT sout,T2400,(#no1)
        SEROUT sout,T2400,(cr,lf,"Second number:")
        SERIN sin,T2400,#no2
        SEROUT sout,T2400,(#no2)
   '
   ' Get the required operation
   '
        SEROUT sout,T2400,(cr,lf,"Operation:")
        SERIN sin,T2400,oper
        SEROUT sout,T2400,(oper,cr,lf)
'
' Now perform the required operation
'
        IF oper = "+" THEN add
        IF oper = "-" THEN subtract
        IF oper = "*" THEN multiply
        IF oper = "/" THEN divide
        GOTO loop


'
' Add the 2 numbers
'
add:
        result = no1 + no2
        GOTO disply


'
' Subtract the 2 numbers
'
subtract:
        result = no1 - no2
        GOTO disply
'
' Multiply the 2 numbers
'
multiply:
        result = no1 * no2
        GOTO disply
'
' Divide the 2 numbers
'
divide:
        result = no1 / no2
'
' Now send the result to the RS-232 port
'
disply:
        SEROUT sout,T2400,("Result=",#result)
        GOTO loop

        END
```

Fig. 8.10 (Continued)

and the user is prompted to enter a new password. This password is stored in addres 0 using the command:

WRITE 0, pwd

where variable *pwd* contains the character entered by the user. If this is not the firs time, i.e. a password was entered before, then address 0 contains data other than 25! and the program jumps to label *chk_pwd*. The user is prompted to enter the password and this is checked with the one in the EEPROM. If the password is correct the program continues. Otherwise, the program jumps back to label *chk_pwd* and prompts the user to re-enter the password again.

PROJECT 29 – Output to a Serial LCD

PIC BASIC language does not have any built-in instructions to send data directly to a standard parallel LCD display. As a result of this it is quite complex to drive an LCD using the PIC BASIC language and this is beyond the scope of this book. It is, however, possible to connect a serial LCD to a PIC microcontroller and then send data to this LCD using the serial output command SEROUT. This and the following section describe two projects which show how to use serial LCDs from the PIC BASIC language.

An ILM-216 type serial LCD module is used in this and the following project. This is a 16-pin, 2-line by 16-character LCD unit manufactured by Scott Edwards Electronics Inc. The device can operate with a baud rate from 1200 to 9600. In addition to the normal display functions, inputs for four push-button switches and also an output to drive a buzzer are included on the LCD circuit. The device incorporates an EEPROM memory and a backlight which are programmable.

Figure 8.11 is a picture of the ILM-216 LCD module. The device has 16 pins and Table 8.5 shows the pin configuration and the functions of these pins. Pins 1 and 2 are the ground and the +5 V supply connections respectively. Pin 3 is the serial input pin. Either RS232 voltage levels or standard TTL level signals can be connected to this pin. Similarly, pin 4 is the serial output pin and TTL logic levels (inverted) can be connected to this pin. Pin 5 is the bell out pin where a small buzzer (up to 25 mA) can be connected to this pin and the buzzer can be controlled from software. Pins 6, 7 and 8 are the option pins. Pin 7 is used to configure the device. Pin 8 is used to select a baud rate and when this pin is connected to pin 6 the device operates at 9600 baud. Leaving pin 8 unconnected configures the device to operate at 2400 baud. Pins 9 to 16 are four push-button switch inputs. The state of these pins can be read from software.

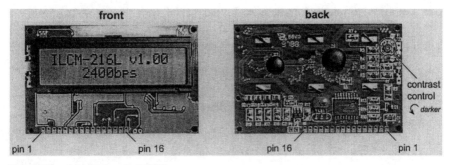

Fig. 8.11 ILM-216 serial LCD

Table 8.5 Pin configuration of ILM-216

Pin	Function
1	GND
2	+5 V
3	Serial in
4	Serial out
5	Bell
6	GND
7	Config/test
8	9600 baud
9	Switch 1
10	Switch 1 GND
11	Switch 2
12	Switch 2 GND
13	Switch 3
14	Switch 3 GND
15	Switch 4
16	Switch 4 GND

The ILM-216 can be connected to a microcontroller using the following minimum pins:

Pin 1   ground

Pin 2   +5 V supply

Pin 3   to microcontroller serial output

Pin 4   to microcontroller serial input

The default configuration of the ILM-216 is 2400 baud, 8 data bits, no parity, and 1 stop bit. The ILM-216 is controlled from its serial input. Table 8.6 gives a list of the control codes and these are summarized below:

Null (ASCII code 0): These characters are ignored by the LCD.

Cursor home (ASCII code 1): Moves the cursor to the first character position of the first line.

Hide cursor (ASCII code 4): Hides the cursor so that it is not visible.

Table 8.6 ILM-216 LCD control codes

Function	ASCII Code
Null	0
Cursor home	1
Hide cursor	4
Show underline cursor	5
Show blinking cursor	6
Bell	7
Backspace	8
Horizontal tab	9
Smart line feed	10
Vertical tab	11
Clear screen	12
Carriage return	13
Backlight on	14
Backlight off	15
Cursor position	16
Format right-aligned text	18
Escape codes	27

Show underline cursor (ASCII code 5): Shows a non-blinking underlined cursor at the current position.

Show blinking cursor (ASCII code 6): Shows a blinking cursor at the current cursor position.

Bell (ASCII code 7): Sends pulses to a buzzer connected to pin 5 of the LCD.

Backspace (ASCII code 8): Moves the cursor back one space and erases the character in that position.

Smart line feed (ASCII code 10): Moves the cursor down one line.

Vertical tab (ASCII code 11): Moves the cursor up one line.

Clear screen (ASCII code 12): Clears the LCD screen.

Carriage return (ASCII code 13): Moves the cursor to the first position of the next line.

Backlight ON (ASCII code 14): Turns on the LED backlight.

Backlight OFF (ASCII code 15): Turns off the LED backlight.

Position cursor (ASCII code 16): Accepts a number from 0 to 31 and moves the cursor to that screen position where 0 is the first character of the first line and 31 is the last character of the second line. Number 64 should be added to the required cursor position in order to get the actual displayed cursor position. For example, position 80 corresponds to the first character position in the second line (64 + 16 = 80).

Right align text (ASCII code 18): Accepts a number from 2 to 9 representing the width of an area on the screen in which right-aligned text is to be displayed.

Escape sequences (ASCII code 27): Escape codes enable the user to define a custom character, to transfer data from the EEPROM, and to read the state of the four push-button switch positions on the LCD module.

Function

This project shows how we can display a two-line message on the ILM-216 serial display. The message "LCD LINE 1" is displayed in line 1 and the message "LCD LINE 2" is displayed in line 2 of the LCD.

Circuit Diagram

Figure 8.12 shows the complete circuit diagram of Project 29. A PIC16F84 type microcontroller is used in this project. Pin 0 of port B is used as the serial output port

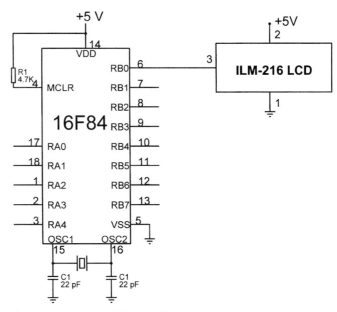

Fig. 8.12 Circuit diagram of Project 29

and is directly connected to pin 3 of the LCD. Since we are not using the other features of the LCD, there is no need to make any other connections between the microcontroller and the LCD.

Program Description

The program simply sends out serial data to display a message on the LCD. The following PDL describes the operation of the program:

START

> Wait a second until the LCD is ready
>
> Clear the LCD
>
> Send message "LCD LINE 1" to the first line
>
> Send message "LCD LINE 2" to the second line

END

Program Listing

The complete program listing is given in Fig. 8.13. Symbol sout is assigned to be the serial output pin. The program delays for a second for the LCD to initialize and then the LCD is cleared. The message "LCD LINE 1" is then sent to port sout which is displayed in line 1 of the LCD. Then the message "LCD LINE 2" is sent to line 2 of the display. Note that, as shown in Fig. 8.14, the second line starts from character position 16 and 64 is added to this position to get the first character position of the second line.

```
'****************************************************************
'
'           PROJECT:       PROJECT29
'           FILE:          PROJ29.BAS
'           DATE:          August 2000
'           PROCESSOR:     P16F84
'           COMPILER:      PIC BASIC
'
'
' This project sends the message LCD LINE 1 to the first line
' of the LCD, and the message LCD LINE 2 is then sent to the second
' line of the LCD.  The LCD is cleared before the message is sent.
'****************************************************************

symbol sout = 0
symbol sec = 1000

        PAUSE sec

        SEROUT sout,N2400,(12,"LCD LINE 1")
        SEROUT sout,N2400,(16,80,"LCD LINE 2")

loop:   GOTO loop
        END
```

Fig. 8.13   Program listing of Project 29

Actual LCD character positions

0	1	2	3	4	5	6	7	8	9	10	11	12	13	14	15
16	17	18	19	20	21	22	23	24	25	26	27	28	29	30	31

Data to send with ASCII code 16 to move to a position

64	65	66	67	68	69	70	71	72	73	74	75	76	77	78	79
80	81	82	83	84	85	86	87	88	89	90	91	92	93	94	95

Fig. 8.14 LCD cursor positions

PROJECT 30 – Storing and Displaying the Data in EEPROM

Function

This project shows how we can store data in the EEPROM of a PIC16F84 microcontroller and then display this data on a serial LCD. The data "1234567890" is stored starting from address 0 of the EEPROM. The first 10 locations of the EEPROM are then read and displayed on an ILM-216 type serial LCD in the following format:

FIRST 10 EEPROM:

1234567890

Circuit Diagram

The circuit diagram of this project is the same as in Fig. 8.12.

Program Description

The following PDL describes the operation of the program:

START

Wait a second until the LCD is ready

Clear LCD

Store data "1234567890" in the first 10 EEPROM locations

Read the contents of the first 10 EEPROM locations

Display the data on a serial LCD

END

Program Listing

The complete program listing of Project 30 is given in Fig. 8.15. Data is stored in the first 10 locations of the EEPROM using the command:

EEPROM 0, (1,2,3,4,5,6,7,8,9,0)

A SEROUT command is then used to clear the LCD and display the message "FIRST 10 EEPROM:" on the first line of the LCD. A FOR loop is used together with a READ command to read the contents of the first 10 EEPROM locations.

The data is sent to the LCD, starting from the first character position of line 2 (i.e. location 64 + 16 = 80). Note that variable *adr* is incremented to move to the next character position after a data value is sent to the display. The LCD will display the following data:

FIRST 10 EEPROM:

1234567890

```
'****************************************************************
'
'           PROJECT:        PROJECT30
'           FILE:           PROJ30.BAS
'           DATE:           August 2000
'           PROCESSOR:      P16F84
'           COMPILER:       PIC BASIC
'
'
' This project stores the data 1234567890 in the first 10
' locations of the EEPROM.  The following data is then displayed
' on the ILM-216 serial LCD:
'
'           FIRST 10 EEPROM:
'           1234567890
'
'****************************************************************
symbol eprm = B0
symbol cnt = B1
symbol adr = B2
symbol sout = 0
symbol sec = 1000

        PAUSE sec                               'Wait for a second

        EEPROM 0,(1,2,3,4,5,6,7,8,9,0)          'Store data in EEPROM
'
        SEROUT sout,N2400,(12,"FIRST 10 EEPROM:")
'
' Now read the first 10 locations of EEPROM and display
' the data on the LCD.
'
        FOR cnt = 0 TO 9
                READ cnt,eprm
                adr = 80 + cnt
                SEROUT sout,N2400,(16,adr,#eprm)
        NEXT cnt
'
' Wait here forever
'
loop:   GOTO loop
        END
```

Fig. 8.15

Appendix A

ASCII Code

ASCII codes of the first 128 characters are standard and the same code is used between different equipment manufacturers. ASCII codes of characters between 128 and 255 are also known as the extended ASCII characters and these characters and their codes may differ between different computer manufacturers. Below is a list of the most commonly used ASCII characters and their codes both in hexadecimal and in binary.

Character	Hex	Binary	Character	Hex	Binary
NUL	00	00000000	DC2	12	00010010
SOH	01	00010001	XOFF	13	00010011
STX	02	00100010	DC4	14	00010100
ETX	03	00110011	NAK	15	00011001
EOT	04	01000100	SYN	16	00010110
ENQ	05	10011001	ETB	17	00010111
ACK	06	01100110	CAN	18	00011000
BEL	07	01110111	EM	19	00011001
BS	08	10001000	SUB	1A	00011010
HT	09	10011001	ESC	1B	00011011
LF	0A	10101010	FS	1C	00011100
VT	0B	10111011	GS	1D	00011101
FF	0C	11001100	RS	1E	00011110
CR	0D	11011101	US	1F	00011111
SO	0E	11101110	SP	20	00100000
SI	0F	11111111	!	21	00100001
DLE	10	00010000	"	22	00100010
XON	11	00010001	#	23	00100011

Character	Hex	Binary	Character	Hex	Binary
$	24	00100100	B	42	01000010
%	25	00101001	C	43	01000011
&	26	00100110	D	44	01000100
'	27	00100111	E	45	01001001
(	28	00101000	F	46	01000110
)	29	00101001	G	47	01000111
*	2A	00101010	H	48	01001000
+	2B	00101011	I	49	01001001
,	2C	00101100	J	4A	01001010
–	2D	00101101	K	4B	01001011
.	2E	00101110	L	4C	01001100
/	2F	00101111	M	4D	01001101
0	30	00110000	N	4E	01001110
1	31	00110001	O	4F	01001111
2	32	00110010	P	50	10010000
3	33	00110011	Q	51	10010001
4	34	00110100	R	52	10010010
5	35	00111001	S	53	10010011
6	36	00110110	T	54	10010100
7	37	00110111	U	55	10011001
8	38	00111000	V	56	10010110
9	39	00111001	W	57	10010111
:	3A	00111010	X	58	10011000
;	3B	00111011	Y	59	10011001
<	3C	00111100	Z	5A	10011010
=	3D	00111101	[	5B	10011011
>	3E	00111110	\	5C	10011100
?	3F	00111111	]	5D	10011101
@	40	01000000	^	5E	10011110
A	41	01000001	_	5F	10011111

Character	Hex	Binary	Character	Hex	Binary
`	60	01100000	~	7E	01111110
a	61	01100001		7F	01111111
b	62	01100010		80	10000000
c	63	01100011		81	10000001
d	64	01100100	,	82	10000010
e	65	01101001	ƒ	83	10000011
f	66	01100110	„	84	10000100
g	67	01100111	. . .	85	10001001
h	68	01101000	†	86	10000110
i	69	01101001	‡	87	10000111
j	6A	01101010	^	88	10001000
k	6B	01101011	‰	89	10001001
l	6C	01101100	Š	8A	10001010
m	6D	01101101	‹	8B	10001011
n	6E	01101110	œ	8C	10001100
o	6F	01101111		8D	10001101
p	70	01110000		8E	10001110
q	71	01110001		8F	10001111
r	72	01110010		90	10010000
s	73	01110011	'	91	10010001
t	74	01110100	'	92	10010010
u	75	01111001	"	93	10010011
v	76	01110110	"	94	10010100
w	77	01110111	•	95	10011001
x	78	01111000	–	96	10010110
y	79	01111001	‾	97	10010111
z	7A	01111010	~	98	10011000
{	7B	01111011	(tm)	99	10011001
|	7C	01111100	š	9A	10011010
}	7D	01111101	›	9B	10011011

Character	Hex	Binary	Character	Hex	Binary
œ	9C	10011100	°	BA	10111010
	9D	10011101	»	BB	10111011
	9E	10011110	1/4	BC	10111100
Ÿ	9F	10011111	1/2	BD	10111101
	A0	10100000	3/4	BE	10111110
¡	A1	10100001	¿	BF	10111111
¢	A2	10100010	À	C0	11000000
£	A3	10100011	Á	C1	11000001
¤	A4	10100100	Â	C2	11000010
¥	A5	10101001	Ã	C3	11000011
¦	A6	10100110	Ä	C4	11000100
§	A7	10100111	Å	C5	11001001
¨	A8	10101000	Æ	C6	11000110
(c)	A9	10101001	Ç	C7	11000111
ª	AA	10101010	È	C8	11001000
«	AB	10101011	É	C9	11001001
¬	AC	10101100	Ê	CA	11001010
–	AD	10101101	Ë	CB	11001011
(r)	AE	10101110	Ì	CC	11001100
¯	AF	10101111	Í	CD	11001101
°	B0	10110000	Î	CE	11001110
±	B1	10110001	Ï	CF	11001111
²	B2	10110010	Ð	D0	11010000
³	B3	10110011	Ñ	D1	11010001
´	B4	10110100	Ò	D2	11010010
µ	B5	10111001	Ó	D3	11010011
¶	B6	10110110	Ô	D4	11010100
·	B7	10110111	Õ	D5	11011001
¸	B8	10111000	Ö	D6	11010110
¹	B9	10111001	×	D7	11010111

Character	Hex	Binary	Character	Hex	Binary
Ø	D8	11011000	ì	EC	11101100
Ù	D9	11011001	í	ED	11101101
Ú	DA	11011010	î	EE	11101110
Û	DB	11011011	ï	EF	11101111
Ü	DC	11011100	ð	F0	11110000
Ý	DD	11011101	ñ	F1	11110001
Þ	DE	11011110	ò	F2	11110010
ß	DF	11011111	ó	F3	11110011
à	E0	11100000	ô	F4	11110100
á	E1	11100001	õ	F5	11111001
â	E2	11100010	ö	F6	11110110
ã	E3	11100011	÷	F7	11110111
ä	E4	11100100	ø	F8	11111000
å	E5	11101001	ù	F9	11111001
æ	E6	11100110	ú	FA	11111010
ç	E7	11100111	û	FB	11111011
è	E8	11101000	ü	FC	11111100
é	E9	11101001	ý	FD	11111101
ê	EA	11101010	Þ	FE	11111110
ë	EB	11101011	ÿ	FF	11111111

Appendix B

PIC BASIC Instruction Set

ASM ... ENDASM	Insert assembly language code
BRANCH	Computed GOTO
BUTTON	Debounce and auto-repeat on specified pin
CALL	Call assembly language subroutine
EEPROM	Store data in EEPROM
END	Stop execution and enter low power mode
FOR ... NEXT	Execute a number of times
GOSUB	Call BASIC subroutine
GOTO	Jump to a specified label
HIGH	Make a pin logic HIGH
I2CIN	Read bytes from an I^2C device
I2COUT	Send out data to an I^2C device
IF ... THEN	Conditional execution of statements
INPUT	Make a pin input
LET	Assign to a variable
LOOKDOWN	Search a table for a value
LOOKUP	Fetch value from a table
LOW	Make a pin logic LOW
NAP	Power down the processor
OUTPUT	Make a pin output
PAUSE	Delay specified number of milliseconds
PEEK	Read data from a register
POKE	Store data in a register
POT	Read potentiometer on a specified pin

PULSIN	Measure pulse width on a pin
PULSOUT	Send a pulse to a specified pin
PWM	Output pulse width modulated data
RANDOM	Generate a random number
READ	Read byte from on-chip EEPROM
RETURN	Return from a subroutine
REVERSE	Make output pin an input or an input pin an output
SERIN	Read serial data in
SEROUT	Send serial data out
SLEEP	Power down the processor for a period of time
SOUND	Generate tone on a specified pin
TOGGLE	Toggle the state of an output pin
WRITE	Write data byte to the on-chip EEPROM

Appendix C

PIC BASIC PRO Instruction Set

@	Insert a line of assembly language code
ADCIN	Read on-chip A/D converter
ASM ... ENDASM	Insert assembly language code
BRANCH	Computed GOTO
BRANCHL	Long BRANCH
BUTTON	Debounce and auto-repeat on specified pin
CALL	Call assembly language subroutine
CLEAR	Zero all variables
CLEARWDT	Clear watchdog timer
COUNT	Count number of pulses on a pin
DATA	Store data in on-chip EEPROM
DEBUG	Serial data out to a fixed pin
DEBUGIN	Serial input from a fixed pin
DISABLE	Disable interrupt processing
DISABLE DEBUG	Disable ON DEBUG processing
DISABLE INTERRUPT	Disable ON INTERRUPT processing
DTMFOUT	Generate touch tones on a specified pin
EEPROM	Store data in EEPROM
ENABLE	Enable ON INTERRUPT and ON DEBUG processing
ENABLE DEBUG	Enable ON DEBUG processing
ENABLE INTERRUPT	Enable ON INTERRUPT processing
END	Stop execution and enter low power mode
FOR ... NEXT	Execute a number of times

FREQOUT	Generate frequencies on a specified pin
GOSUB	Call BASIC subroutine
GOTO	Jump to a specified label
HIGH	Make a pin logic HIGH
HSERIN	Hardware serial input
HSEROUT	Hardware serial output
I2CREAD	Read bytes from an I^2C device
I2CWRITE	Write bytes to an I^2C device
IF ... THEN ... ELSE	Conditional execution of statements
INPUT	Make a pin input
LCDIN	Read RAM on an LCD
LCDOUT	Display data on an LCD
LET	Assign to a variable
LOOKDOWN	Search a table for a value
LOOKDOWN2	Search variable table for a value
LOOKUP	Fetch value from a table
LOOKUP2	Fetch variable from a table
LOW	Make a pin logic LOW
NAP	Power down the processor
ON DEBUG	Execute debug monitor
ON INTERRUPT	Execute a specified code on an interrupt
OUTPUT	Make a pin output
PAUSE	Delay specified number of milliseconds
PAUSEUS	Delay specified number of microseconds
PEEK	Read data from a register
POKE	Store data in a register
POT	Read potentiometer on a specified pin
PULSIN	Measure pulse width on a pin
PULSOUT	Send a pulse to a specified pin
PWM	Output pulse width modulated data
RANDOM	Generate a random number

RCTIME	Measure pulse width on a specified pin
READ	Read byte from on-chip EEPROM
RESUME	Continue normal execution after an interrupt
RETURN	Return from a subroutine
REVERSE	Make output pin an input or an input pin an output
SERIN	Read serial data in
SERIN2	Read serial data in
SEROUT	Send serial data out
SEROUT2	Send serial data out
SHIFTIN	Receive synchronous serial data
SHIFTOUT	Send synchronous serial data out
SLEEP	Power down the processor for a period of time
SOUND	Generate tone on a specified pin
STOP	Stop program execution
SWAP	Exchange the values of two variables
TOGGLE	Toggle the state of an output pin
WHILE ... WEND	Exececute code enclosed while the condition is true
WRITE	Write data byte to the on-chip EEPROM
XIN	X-10 input
XOUT	X-10 output

Appendix D

LET BASIC Instruction Set

ADIN	Retrieve the A/D value on the PIC16C71
ASM	Insert assembly code in BASIC
BSTART	I²C bus interface
BSTOP	I²C bus interface
BUSIN	Get a byte from the I²C BUS
BUSOUT	Output a number to the I²C BUS
BUTTON	Wait for a pin to invert its state
CLEAR	Set a pin to logic LOW state
CLS	Clear the LCD and home the cursor
COUNTER	Clear and enable a counter
CURSOR	Move the LCD cursor to a position
DATA	Define a table of alphanumeric data
DEFINE	Define port pins as inputs or outputs
DELAYMS	Delay program execution specified number of milliseconds
DELAYUS	Delay program execution specified number of microseconds
DEVICE	Specify the type of microcontroller used
DIM	Declare a variable
EEDATA	Read data from internal EEPROM of 16F84 or 16C84
END	Stop compilation of source code
FOR ... NEXT	Execute a number of times
GOSUB	Go to a subroutine
GOTO	Jumps to the specified label
IF ... THEN	Conditional execution
INCLUDE	Include predefined packages in the compilation

INIT	Assign code from predefined packages
INKEY	Wait for a key press
INPORTA	Get data from port A
INPORTB	Get data from port B
INPORTC	Get data from port C
LET	Assign an expression to variable
MEMREAD	Read from I^2C bus device
MEMWRITE	Write to I^2C bus device
OUTA	Send data to port A
OUTB	Send data to port B
OUTC	Send data to port C
PEEK	Get value from a register
POKE	Store value in a register
PRINT	Display data on the LCD
READ	Read the next data item from a DATA table
REM	Comment in a program
RESTORE	Change the position of the pointer in a DATA table
RSIN	Serial data in
RSOUT	Serial data out
SET	Set a pin into logic HIGH state
SLEEP	Place the microcontroller in power down mode
SOUND	Send a tone to a specified pin
STOP	Halt the program execution
STORE	Store data in the EEPROM memory
SWAP	Swap two variables
SYMBOL	Assign a name to a port pin
TIMER	Clear and enable the internal timer

Glossary

ADC   Analogue-to-digital Converter. A device that converts analogue signals to a digital form for use by a computer.

Algorithm   A fixed step-by-step procedure for finding a solution to a problem.

ANSI   American National Standards Institute.

Architecture   The arrangement of functional blocks in a computer system.

Array   A variable with multiple elements of the same size.

ASCII   American Standard Code for Information Interchange. A widely used code in which alphanumeric characters and certain other special characters are represented by unique 7-bit binary numbers. For example, the ASCII code of letter "A" is 65.

Assembler   A software that translates symbolically represented instructions into their binary equivalents.

Assembly language   A source language which is made up of the symbolic machine language statements. Assembly language is very efficient since there is a one-to-one correspondence with the instruction formats and data formats of the computer.

BASIC   Beginners All-purpose Symbolic Instruction Code. A high-level programming language commonly used in personal computers. BASIC is usually an interpreted language.

Baud   The unit of data transmission speed. Baud is often equated to the number of serial bits transferred per second.

Baud Rate   Measurement of data flow in a serial communication system. Baud rate is typically equal to bits per second. Some typical baud rates are 9600, 4800, 2400 and so on.

BCD   Binary Coded Decimal. A code in which each decimal digit is binary coded into 4-bit words. By representing binary numbers in this form, it is readily possible to display and print numbers.

Bidirectional port   An interface port that can be used to transfer data in either direction.

Binary The representation of numbers in base two system.

Bit A single binary digit.

Bug An error in a program.

Byte A group of 8 binary digits.

Chip A small rectangle of silicon on which an integrated circuit is fabricated.

Clock A circuit generating regular timing signals for a digital logic system. In microcomputer systems clocks are usually generated by using crystal devices. A typical clock frequency is 12 MHz.

CMOS Complementary Metal Oxide Semiconductor. A family of integrated circuits that offers extremely high packing density and low power.

Compiler A program designed to translate high-level languages into machine code.

Complement Changing all the bits to opposite state.

Constant A number that cannot change during the execution of a program.

Counter A register or a memory location used to record numbers of events as they occur.

CRT Cathode Ray Tube. A display screen.

Cycle time Time required to access a memory location or to carry out an operation in a computer system.

DAC Digital to Analogue Converter. A device that converts digital signals into analogue form.

Decimal system Base 10 numbering system.

Development system Equipment used to develop microprocessor- and micro-computer-based software and hardware projects.

Dot matrix Method of printing or displaying characters in which each character is formed by a rectangular array of dots to give the required shape.

EAROM Electrically Alterable Read Only Memory. In this type of memory part or all of the data can be erased and rewritten by applying electrical signals.

Edge triggered Circuit action is initiated by the change of a signal. An edge could be the change of a signal from 0 to 1 or from 1 to 0.

EEPROM Electrically Erasable, Programmable, Read-Only Memory. Data stored in such a memory is permanent and is not lost when the power is removed. EEPROM memory contents can be changed, replacing old data with new.

Emulator Software or hardware system that duplicates the actions of a microprocessor or a microcomputer system.

EPROM   Erasable Programmable Read Only Memory. This type of memory can be erased by exposure to ultraviolet light and then reprogrammed using a programmer.

Execute   To perform a specified operational sequence in a program.

File   Logical collection of data.

Flowchart   Graphical representation of the operation of a program.

Gate   A logic circuit having one or more inputs and a single output. For example, NAND gate.

Half duplex   A two-way communication system that permits communication in one direction at a time.

Hardware   The physical parts or electronic circuitry of a computer system.

Hexadecimal   Base 16 numbering system. In hexadecimal notation, numbers are represented by digits 0–9 and the characters A–F. For example, decimal number 165 is represented as A5.

High-level language   Programming language in which each instruction or statement corresponds to several machine code instructions. Some high-level languages are BASIC, FORTRAN, C, PASCAL and so on.

Input device   An external device connected to the input port of a computer. For example, a keyboard is an input device.

Input port   Part of a computer that passes external signals into a computer. Microcomputer input ports are usually 8 bits wide.

I/O   Short for Input/Output.

Input/Output   The hardware within the computer that connects the computer to external peripherals and devices.

Instruction cycle   The process of fetching an instruction from memory and executing it.

Instruction set   The complete set of instructions of a microprocessor or a microcomputer.

Integer   Whole number with no fractional part.

Interface   To interconnect a computer to external devices and circuits.

Interrupt   An external or internal event that suspends the normal program flow within a computer and causes entry into a special interrupt program (also called the interrupt service routine). For example, an external interrupt could be generated when a button is pressed. An internal interrupt could be generated when a timer reaches a certain value.

Interrupt vector   Reserved memory locations where a program jumps when an interrupt is detected.

ISR   Interrupt Service Routine. A program that is entered when an external or an internal interrupt occurs. Interrupt service routines are usually high-priority routines.

K   Multiplier for 1024. For example, 1 K bytes is 1024 bytes.

Language   A prescribed set of characters and symbols which is used to convey a program to a computer.

LCD   Liquid Crystal Display. A low-powered display which operates on the principle of reflecting incident light. An LCD does not itself emit light. There are many varieties of LCDs. For example, numeric, alphanumeric, or graphical.

LED   Light Emitting Diode. A semiconductor device that emits a light when a current is passed in the forward direction. There are many colours of LEDs. For example, red, yellow, green, and white.

Level triggered   Circuit action is initiated by the presence of a signal.

Logic levels   Voltage levels representing the two logical states (0 and 1) of a digital signal. Logic HIGH is also called state 1 and logic LOW is called state 0.

Loop   Sequence of program code that is executed repeatedly.

Machine code   Lowest level in which programs are written. Machine code is usually written in hexadecimal.

Microcomputer   General purpose computer using a microprocessor as the CPU. A microcomputer consists of a microprocessor, memory, and input/output.

Microprocessor   A single large-scale integrated circuit which performs the functions of a CPU.

Mnemonic   A programming short-hand using letters, numbers and symbols adopted by each manufacturer to represent the instruction set of a microprocessor.

Nibble   A group of 4 binary bits.

NMOS   Negative channel Metal Oxide Semiconductor. A device based on n-channel field-effect transistor cell.

Non-volatile memory   A semiconductor memory type that holds data even if power has been disconnected.

Octal   Representation of numbers in base 8.

Op-code   Operation code. That part of an instruction which specifies the function to be performed.

Output device   An external device connected to the output port of a computer. For example, a printer is an output device.

Output port   Part of a computer which is used to pass electrical signals to outside the computer. Microcomputer output ports are usually 8 bits wide.

Parity   A binary digit added to the end of an array of bits to make the sum of all ones either odd or even. Parity is a method of checking the accuracy of transmitted or received binary data.

PDL   Program Description Language. Representation of the control and data flow in a program using simple English-like sentences.

PEROM   Flash Programmable and Erasable Memory. This type of memory can be erased and reprogrammed using electrical signals only, i.e. there is no need to use an ultraviolet light source to erase the memory.

Port   An electrical logic circuit that is a signal input or output access point of a computer.

Programmed I/O   The control of data flow in and out of a computer under software control.

PROM   Programmable Read Only Memory. A type of semiconductor memory which can be programmed by the user using a special piece of equipment called a PROM programmer (or PROM blower).

Pull down resistor   A resistor connected between an I/O pin and ground.

Pull up resistor   A resistor connected to the output of an open collector (or open drain) transistor of a gate in order to load the output.

RAM   Random Access Memory, also called read/write memory. Data in RAM is said to be volatile and it is present only as long as the chips have power supplied to them. When the power is cut off, all information disappears.

Register   A storage element in a computer. A register is usually 8 bits wide in most microprocessors and microcomputers.

ROM   Read Only Memory. A type of semiconductor memory which is read only.

RS232   An internationally recognized specification for serial data transfer between two devices.

Serial   Information transfer on a single wire where each bit is transferred sequentially with a time delay in between.

Software   Program.

Source code   The human readable version of a computer program.

Start bit   The first bit sent in a serial communication. There is only one start bit in a frame of serial communication.

Stop bit   The last bit sent in a serial communication. There can be 1 or 2 stop bits per frame of a serial communication.

Syntax   The rules governing the structure of a programming language.

Transducer   A device that converts a measurable quantity into an electronic signal. For example, a temperature transducer gives out an electrical signal which may be proportional to the temperature.

TTL   Transistor Transistor Logic. A kind of bipolar digital circuit.

UART   Universal Asynchronous Receiver Transmitter. This is a semiconductor chip which converts parallel data into serial form and serial data into parallel form. A UART is used in RS232 type serial communication.

Unsigned number   A number whose sign (positive or negative) is not stored.

USART   Synchronous version of UART.

UV   Ultraviolet light. Used to erase EPROM memories.

Variable   A temporary data storage location in RAM. The contents of variables can be changed in a program.

VDU   Visual Display Unit.

Volatile   Memory that loses its contents when power is lost (e.g. RAM).

Word   A group of 16 binary digits.

Index

209

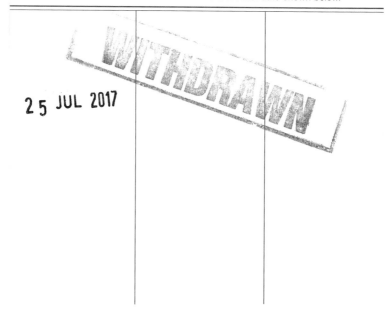